Materials and Structures for Energy Absorption

IMechE Seminar Publication

I MECH E

Materials and Structures for Energy Absorption

Organized by
the Materials and Mechanics of Solids Group of
The Institution of Mechanical Engineers (IMechE)

IMechE Seminar Publication 2000–21

Professional Engineering Publishing

Published by Professional Engineering Publishing Limited for The Institution of Mechanical Engineers, Bury St Edmunds and London, UK.

First Published 2000

ISSN 1357–9193
ISBN 1 86058 321 0

A CIP catalogue record for this book is available from the British Library.

Printed by The Cromwell Press, Trowbridge, Wiltshire, UK.

Organizing Committee

Mr A Harrison (Joint Chairman)
Ford Motor Company Limited, UK

Professor N Jones (Joint Chairman)
University of Liverpool, UK

Papers presented at a one-day seminar *Materials and Structures for Energy Absorption* on 9 May 2000, held at IMechE Headquarters, London, UK

Related Titles of Interest

Title	Editor/Author	ISBN
IMechE Engineers' Data Book – Second Edition	C Matthews	1 86058 248 6
Foreign Object Impact and Energy Absorbing Structure	IMechE Seminar	1 86058 156 0
Impact and Dynamic Fracture of Polymers and Composites (ESIS 19)	J G Williams and A Pavan	0 85298 946 6
Evaluating Materials Properties by Dynamic Testing (ESIS 20)	E van Walle	1 86058 004 1

For the full range of titles published by Professional Engineering Publishing contact:

Sales Department
Professional Engineering Publishing Limited
Northgate Avenue
Bury St Edmunds
Suffolk
IP32 6BW
UK

Tel: +44 (0)1284 724384
Fax: +44 (0)1284 718692

Contents

Vehicle crashworthiness in real world conditions

P A FAY
Ford Motor Company Limited, Basildon, UK

INTRODUCTION

It can be shown that over the last 40 or so years, motor vehicles have become safer. There are two fundamentally different ways in which this has been achieved. The first approach (known as "primary safety") is the development of measures to avoid crashes occurring in the first place. The second approach (known as "secondary safety" or "crashworthiness") is the development of measures to minimise the injuries resulting from a crash once it has happened.

These two approaches are not alternatives. They are complementary but their relative importance has changed over the years and has been much debated. It is however widely believed that improved crashworthiness of vehicles (such as improved occupant protection, improved structural characteristics and advanced restraint systems) can continue to make substantial improvements to vehicle safety [1].

VEHICLE CRASHWORTHINESS

Once a crash has occurred, a number of factors affect the outcome in terms of the level of any injuries sustained.
These include:

- The design of the vehicles involved;
- The road infrastructure (crash barriers, roadside furniture, carriageway separation, etc.);
- The injury tolerance of the individuals involved (often associated with age, gender, build and general health);
- The effectiveness of emergency and medical services.

The fundamental principles of vehicle design for crashworthiness are now well understood. Four basic concepts have been identified [2]:

1. Limiting to tolerable levels the impact forces that are applied to occupants;
2. Providing means for managing the energy of collision while at the same time maintaining adequate survival space for occupants;
3. Containing the occupants within the vehicle survival space during the collision, i.e. minimizing the likelihood of ejection;
4. Protecting the occupants from post crash hazards, primarily fires.

The first of these concepts can be broken down further into three additional principles [3]:

- Occupants should be decelerated over as long a time as possible;
- Occupants should be decelerated over as long a distance as possible;
- The forces decelerating the occupants should be spread over as wide an area of the body as possible and preferably over the stronger skeletal structures.

Using these concepts and principles has led to improved vehicle structures and occupant restraint systems.

CRASHWORTHINESS DEVELOPMENTS – STRUCTURES

Vehicle structural design effort has focused on the crashworthiness in the main impact modes seen in real world accidents: front, rear and side impacts and rollovers.

In frontal impact, vehicle structures have been developed to give improved energy absorption and occupant compartment stability in a wide range of full frontal and offset or angled frontal impact modes over a range of impact speeds [4]. Structures have also being designed to perform well in impacts with narrow objects (such as poles and trees).

In rear impact, attention has focused on providing protection against whiplash injuries, protection for the fuel system and the ability to open doors after the impact [5].

Improving side impact protection has been achieved by improved structures and padding [6], as well as the development of side impact airbag systems.

Vehicle structures have also been developed to give improved rollover performance through reduced risk of occupant ejection, reduced roof crush and improved interior head impact protection [7].

CRASHWORTHINESS DEVELOPMENTS – OCCUPANT RESTRAINT SYSTEMS

Aside from the developments in vehicle structures, progress has been made in improved occupant restraint systems. Seat belts have been fitted to increasing numbers of cars since the 1960s and the belt systems themselves have become more sophisticated. Automatically locking "inertia reel" belts are now the norm. In addition, "pretensioners", "webbing grabbers" and load limiters have been introduced [8].

Clearly, seat belts are only effective if they are actually used. In the US, where seat belt wearing rates tended to be lower than those of Europe, vehicle manufacturers were required to provide protection to unbelted occupants. This was achieved by a number of measures including driver and passenger airbags, knee bolsters and automatic seatbelts [9]. Although airbags have reportedly saved over 1,000 lives in the US, they have also been implicated in the deaths of about 50 occupants. These were mainly children but short stature female drivers and elderly occupants have also been involved.

Frontal impact airbags, designed to give additional protection against head and chest injuries to belted occupants [10], have since been introduced in Europe.

Side impact airbags are also becoming available. There are a number of different approaches being adopted. The three main types are:

- Seat or door mounted airbags designed to protect chest and/or head [11];
- Inflatable tube systems mounted in the door frame [12];
- Inflatable curtains mounted in the roof rail [13].

Considerable work is being carried out to develop restraint systems further. The main activity is developing "smart" or "tailorable" systems [14] which will determine if, when and how to deploy airbags, pretensioners, etc. Techniques are being developed to determine if seats are occupied, the presence and orientation of any child restraints, the weight of the occupants, the fore/aft position of the seat and the severity of the impact.

FUTURE PRIORITIES

Most of the developments described above are aimed at preventing life-threatening injuries to occupants of cars.

Future developments are expected to place increasing importance on:

- injuries which cause long term disablement or impairment. Particular areas identified include leg injuries [15] and neck injuries [16];
- consideration of the level of injuries sustained by the occupants of the other vehicle involved. This so-called "compatibility" [17] between vehicles is of particular importance when the vehicles are of significantly different weights but geometric and structural compatibility are also being investigated;
- injuries sustained by pedestrians impacted by the front end of cars;
- injuries sustained by the elderly who are becoming an increasingly large proportion of car occupants and who may require different restraint characteristics because of their reduced tolerance to loading.

REFERENCES

[1] Mackay, M, Active and passive safety - current and future perspectives, Proceedings of the 3rd International Conference on Innovation and Reliability in Automotive Design and Testing, Firenze, April 1992.

[2] Paluszny, A and Bakkax, J, State of the art review of automobile structural crashworthiness, American Iron and Steel Institute, 1992.

[3] European Transport Safety Council, Reducing traffic injuries through vehicle safety improvements: The role of car design, Technical Annex C, ETSC, Brussels, 1993.
[4] Stucki, S L and Hollowell, W T, NHTSA's Improved frontal protection research program, Proceedings of 15th Experimental Safety Vehicle Conference, Melbourne, May 1996, pp 192 - 205.
[5] Chirwa, E C, Structural crashworthiness simulation of a rear end collision of a small European car, International Journal of Crashworthiness, Vol 1, No 1, 1996, pp 21 - 35.
[6] Hobbs, C A, The influence of car structures and padding on side impact injuries, Structures, Proceedings of the 12th Experimental Safety Vehicle Conference, Göteborg, May 1989, pp 954 - 962.
[7] Summers, S, Rains, G C and Willke, D T, Current research in rollover and occupant retention, Proceedings of 15th Experimental Safety Vehicle Conference, Melbourne, May 1996, pp 760 - 765.
[8] Mitzkus, J E and Eyrainer, H, Three-point belt improvements for increased occupant protection, SAE Technical Paper, 840395, 1984.
[9] Evans, L, Passive compared to active approaches to reducing occupant fatalities, Proceedings of the 12th Experimental Safety Vehicle Conference, Göteborg, May 1989, pp 1149 - 1156.
[10] Kozyreff, M and Schaper, D, A new compact European driver air bag system, Proceedings of the 13th Experimental Safety Vehicle Conference, Paris, November 1991, pp 1176-1180.
[11] Demmler, A, Chest and torso air bag, Automotive Engineering, Vol 104, No 2, p 54, Feb 1996.
[12] Yaniv, G, Romeo, D, Kompass, K and Mayer, J, ITS - A new restraint system for side impact protection, SAE Technical paper 961018, 1996.
[13] Thomas, C M, Swede touts curtain to increase side impact safety, Automotive News, 3 October 1994, p 41.
[14] Galer, M and Jones, S, Intelligent occupant protection systems, Proceedings of 14th Experimental Safety Vehicle Conference, Munich, May 1994, pp 491-498
[15] Thomas, P, Charles, J and Fay, P, Lower limb injuries - The effect of intrusion, crash severity and the pedals on the risk and injury type in frontal collisions, Proceedings of the 39th Stapp Conference, 1995.
[16] Nilson, G, Neck injuries and concepts, Proceedings of the International Symposium on Real World Crash Injury Research, Hinckley, June 1997.
[17] Hobbs, C A, Williams, D A and Coleman, D J, Compatibility of cars in frontal and side impact, Proceedings of 15th Experimental Safety Vehicle Conference, Melbourne, May 1996, pp 617-624.

S672/003/2000

Woven polypropylene structures

I M WARD and **P J HINE**
IRC in Polymer Science and Technology, University of Leeds, UK

INTRODUCTION

Recent research at Leeds University has seen the development of a new class of 'self reinforced' polypropylene sheet. The essence of the process, termed hot compaction, is to take layers of woven oriented polypropylene tapes, and compact them together under suitable conditions of temperature and pressure to form a consolidated, homogeneous sheet. The mechanism for consolidation is to melt a small proportion of the skin of each tape, which on cooling recrystallises to form a polypropylene matrix to bind the oriented tapes together. Research has shown that there is a usable processing window where sufficient melted material can be created to bind the woven structure together without losing too much of the original oriented structure. The process can be seen as a way of converting the properties of a highly drawn polypropylene tape into a consolidated sheet.

The compacted polypropylene tapes display an interesting portfolio of mechanical properties, reflecting the 'frozen in' molecular orientation, and by virtue of being composed of only a single phase, namely polypropylene. These include reasonable mechanical properties (placed between isotropic polymers and glass reinforced polymers), thermoformability, low density and recyclability. Perhaps the most interesting property of the compacted PP sheets is their excellent impact strength. Figure 1 compares the notched IZOD impact strength (ASTM D256) of compacted PP sheet with other comparative materials.

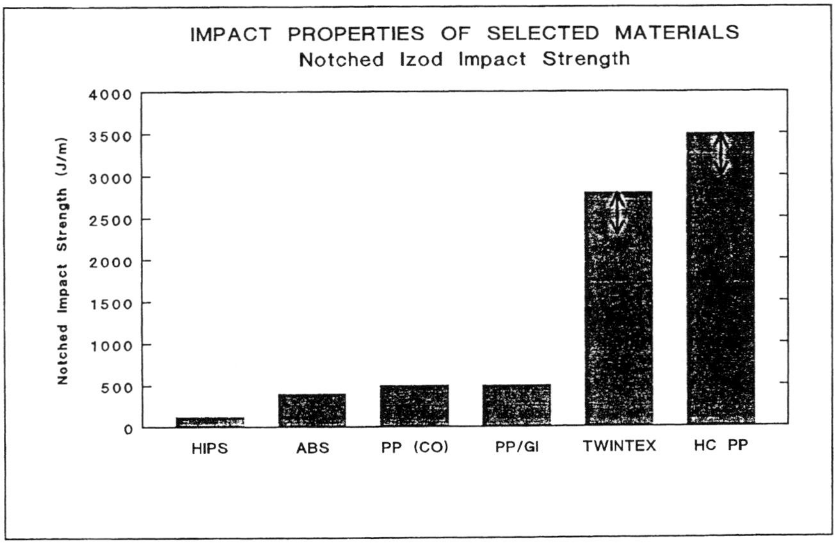

Figure 1 Impact properties of selected materials

HIPS – High impact polystyrene: **ABS** – Acrylonitrile butadiene styrene: **PP (CO)** PP copolymer: **PP/GI** – short glass fibre reinforced PP: **TWINTEX** – woven glass fibre reinforced PP: **HC PP** – hot compacted PP sheet (6mm thick).

It is seen that the notched IZOD impact strength of the hot compacted PP sheet is comparable with woven glass reinforced PP and much higher than most other materials. Interestingly, if we test compacted PP samples in which the fraction of the oriented phase remaining is significantly reduced from that seen in Figure 1, then the notched IZOD strength also falls. This strongly suggests that the origin of the very high impact strengths seen in the compacted PP sheets relates directly to the oriented nature of the original drawn PP tapes.

The geometry of the notched IZOD test is such that starter cracks were placed across the thickness of the sheet, and therefore travelled perpendicular to the direction of preferred molecular orientation. In general, when oriented polymeric structures are impacted perpendicular to the main orientation direction, axial splitting occurs (i.e. transverse to the crack propagation direction), thereby increasing the damage zone and so the amount of energy absorbed. In particular, it is well known that one of the reasons for the high energy absorption of Kevlar fibres is their propensity to split axially. It is likely that the additional internal boundaries in the woven polypropylene sheet, as a result of the woven nature of the starting material, will also aid toughness by acting to blunt the propagating crack tip.

A further interesting aspect of the impact strength of the compacted PP sheets is shown in Table 1, where results are shown for the notched IZOD strength of a particular hot compacted PP sheet, at +20°C and –40°C.

Table 1 Notched IZOD results

Impact energy (J/m) +20°C	Impact energy (J/m) -40°C
2726 ± 240	5225 ± 1228

Against intuition the notched IZOD results at –40°C are actually higher than those at +20°C. The reasons for this are twofold. Firstly polypropylene has a transition in properties around room temperature, due to a mechanical relaxation caused by the structure of the molecular backbone. This results in a value of the bending modulus at –40°C of 7.6GPa, which is almost double the room temperature value. In isotropic polypropylene this leads to brittle behaviour (i.e. lower impact strength), but in the compacted PP sheets where there is preferred molecular orientation, the brittle strength is much enhanced, with the result that at low temperature the material is still ductile. Therefore the combination of an enhanced brittle strength due to molecular orientation, and an increase in modulus, leads to an increase in impact strength as the temperature falls.

Finally we show a table comparing the properties of compacted PP sheets with other materials

Table 2 A comparison table of mechanical properties

	Compacted PP sheet	Isotropic PP (Homopolymer)	Random mat Short glass/PP 40wt% fibre	Twintex Unidirectional 60 wt% fibre
Density (kg/m^{3})	920	900	1190	1500
Tensile Modulus (GPa)	3.5	1.12	8.28	15
Tensile Strength (MPa)	87	27	101	350
Heat deflection temperature (°C)	143	110	154	159
Notched Izod impact strength (J/m)	>3000	200	752	2000
Thermal expansion (/°C)	60	96	27	21

S672/004/2000

Controlling structural collapse in off axis impact of rail vehicles by retrospective modification

P WAITE
WS Atkins Rail Limited, Derby, UK
N KIRK
WS Atkins Consultants Limited, UK

SYNOPSIS

In 1988 British Rail suffered a multi-fatality accident involving the collision and destruction of several vehicles of the Mark 1 type construction, see Figure 1. Vehicle underframes climbing over each other and shearing through the body structural members caused the most severe damage. Subsequent to this accident research and development work was commissioned to establish the feasibility of providing additional strength into these Mark 1 coaches above the underframe and of preventing the underframe structures from climbing over each other during a collision. Whilst strengthening the body structure was found not to be feasible, providing a simple and relatively cheap means of anti-climb was developed and tested on full size vehicles. This paper describes the key steps taken in the design and development of this solution.

INTRODUCTION

In recent years there has been a significant amount of work aimed at improving passenger safety during collisions between vehicles, both in the rail and automotive industries. Previously however, crashworthiness was a less important design consideration, and many old vehicle designs do not meet today's stringent safety standards. In order to improve the safety of older vehicles, modifications can be made to existing structures to improve crash performance. Two important considerations for the design of such modifications are cost and ease of implementation.

In both the rail and automobile industries, the main goal of crashworthy designs is to limit vehicle deformation to areas where there are no passengers, and to absorb the energy of the collision by crush in these regions. In most cases, this means a large amount of deformation is required at the vehicle ends.

Figure 1. Class 423 Electric Multiple Unit (Mk 1 Type Construction)

Another important consideration for crashworthy design, that is particularly relevant to the rail industry, is to prevent one underframe riding over the adjacent underframe. When such overriding occurs with Mark 1 type vehicles, the underframe of a vehicle climbs over the underframe of an adjacent vehicle and then shears through body structural members, destroying the superstructure of the vehicle. Older vehicles are typically constructed from a stiff underframe with a relatively weak superstructure and are therefore particularly prone to overriding during a collision. This occurred for example, in the Clapham Junction accident of 1988 where three Mark 1 trains collided. Anti-climbing devices are required to provide positive engagement between two colliding vehicles such that there is no relative vertical motion and overriding is avoided. In order for the anti-climbing mechanism to work, all components that are located forward of it at the carriage end, such as couplers and gangways, need to collapse backwards into the underframe during a collision.

In line with the above requirements, the modifications discussed in this paper that have been developed for Mark 1 type vehicles, can be divided into three categories: crush initiators, anti-climbing devices and shear-out components. The crush initiators take the form of cutouts in the vehicle underframe. The size and location of these cutouts are chosen to initiate crushing at the vehicle ends, and to promote local rather than global collapse. The anti-climbing device is a 'cup and cone' mechanism which, when it engages, draws the vehicle ends into alignment. In order for the 'cup and cone' to work, a shear-out device to remove the coupler is required. The design work was undertaken by WS Atkins whilst the testing was undertaken by the Health and Safety Laboratory (HSL).

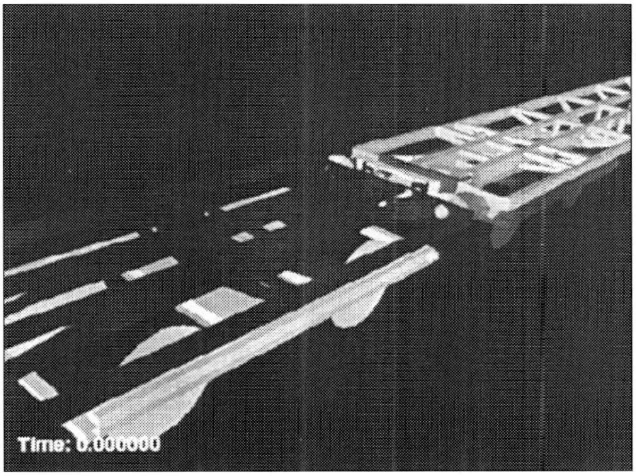

Figure 2. Idealisation of Mk 1 Type Underframe

DESIGN

An isometric idealisation of the Mark 1 vehicle underframe is shown in Figure 2. The underframe has four main longitudinal members. The outer two, referred to as the solebars, are C-channels. Between the headstock and the bolster, the inner members, referred to as the longitudinals, are fabricated from an upper T-section and a lower L-section. There are also two diagonal members, which run between the headstock and the longitudinals which stiffen the drag box area. The upper structure is much weaker than the underframe. The skin is in the most part constructed from 1.6mm steel sheet with a framing structure around windows, doors and panels. The framing is mostly made from 2mm steel sheet pressed into simple angles, but also includes slightly stronger members.

The modifications described in this paper for the Mark 1 type vehicle are all to be located within the vehicle end section between the headstock and the bolster. Each of the three modifications has a series of design requirements and these along with the modification details are summarised below.

Underframe Cutouts

The current design requirement for new vehicles in the UK, set out in the Railway Group Standard GM/RT 2100 (1), stipulates that the energy of a collision must be absorbed first by the vehicle end section before any deformation of the main vehicle body. A minimum of 1 MJ must be absorbed by the end section over a maximum collapse distance of 1.0m, without the force exceeding 3000 kN (a limit of 2500 kN being preferred). The aim of the modifications to the Mark 1 structure is to approach, if not reach, this specification. In order to achieve this aim, a series of underframe cutouts have been developed which selectively weaken the vehicle underframe, initiating collapse at the vehicle end and localising crush in the end section, see Figure 3. It was important for the design that the underframe was not weakened in such a way as to decrease the vertical shear strength of the structure, as this would limit the effectiveness of the anti-overriding devices, or to affect the traction load path through the dragbox.

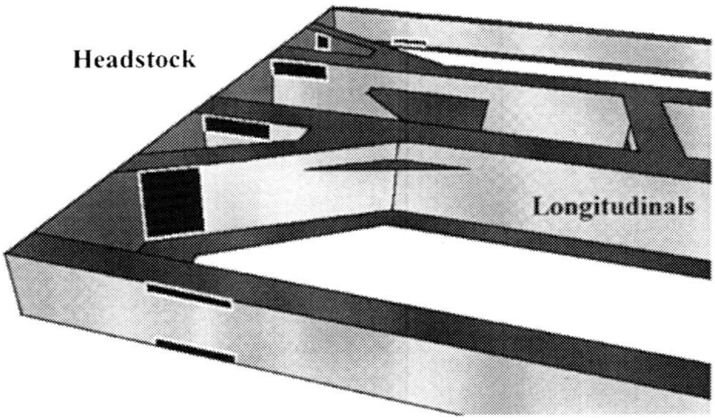

Figure 3. Underframe Crush Initiators

Anti-Climbing Device

The purpose of an anti-climbing device is to provide a strong link between vehicles during a collision. For the particular case of the Mark 1 vehicles, it is necessary that these devices also bring the headstocks into alignment so that the main load path between colliding vehicles is directly through the headstocks to the longitudinal members, thus ensuring the desired underframe collapse modes. The following points must be addressed in the design of such components: (a) sufficient vertical and horizontal shear strength of the component, (b) sufficient vertical and horizontal shear strength of the attachment to the underframe, and (c) engagement over a range of displacement offsets. Displacement offsets can occur for several reasons including suspension dynamics, vehicle end sag, wheel wear, vehicle height adjustment and curving.

The anti-overriding device designed for the Mark 1 modification programme is in the form of a cup and cone, see Figure 4. The design incorporates one cup and one cone attached to each vehicle end such that during a collision between two vehicles, the cone from one vehicle locates in the cup on the facing vehicle, aligning the carriage ends together.

Shear-Out Coupler

Modern designs in the UK incorporate some kind of shear-out device to remove the coupler from the crumple zone during a collision. Normally, this device is composed of a shear-out dragbox, so the entire dragbox and coupler arrangement detach. Unfortunately, modifying the Mark 1 type vehicles to include a shear-out dragbox arrangement would be prohibitively expensive, and therefore an alternative design was required. In the new design, the shear-out section is located within the coupler shank, so it is only the coupler head, which detaches from the main vehicle body.

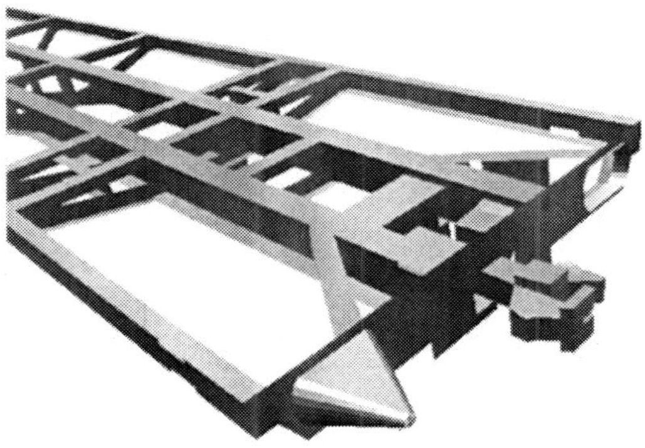

Figure 4. Anti-Climbing Device

There were two main issues to consider during the design of the shear-out coupler: shear-out load and fatigue life. Of these two issues, the fatigue life specification proved the more difficult to address, in the most part due to lack of information concerning the magnitude and number of cycles to which a coupler is subjected during normal use. A conservative estimate of the fatigue loading was therefore used for design purposes.

The shear-out device is composed of two halves of a solid block, which meet on a plane at 45° from the horizontal, see Figure 5. Spanning between the two halves of the block are two dowels near the centre, aligned perpendicular to the plane. There are also four bolts, one at each corner, which are aligned horizontally. The exact geometry of the dowels is dictated by the fatigue performance. The specific bolt grade was chosen for both high yield and tensile strength. The material specified for the dowels not only has both high yield and tensile strength but also low strain to failure, which gives a more easily controlled load at failure.

TESTING ARRANGEMENT

Both static and dynamic tests were carried out on the proposed designs. Static testing was performed on an end section of underframe with the proposed cutouts, see Figure 6. The underframe was tested on a testing frame with a maximum load of 4 MN. The underframe was crushed in seven stages to a final platen displacement of 1000mm.

Dynamic testing consisted of impact testing the cup and cone and coupler components at 40 mph. The impact test arrangement consisted of a 'hammer' truck and a 'brake' truck which run on a hillside track, see Figures 7 & 8. The hammer truck weighs ~ 10 tonnes and runs on wheels. The brake truck weighs ~ 17 tonnes and rests on friction pads on the rails. For the impact testing, the hammer truck is pulled up the hill to a known distance. It is then released, and runs down the hill into the brake truck. A single 'rough shunt' test of the shear-out coupler at 2m/s (~4.5 mph) was also performed. Measurements taken during testing included: vertical and horizontal acceleration of the hammer and brake trucks, impact load and during one of the shear-out coupler tests, the strain in the coupler shank. The impact tests, that were performed, are cup and cone tests with vertical, horizontal, combined vertical and horizontal and no offset, and coupler tests with horizontal, vertical and no offset.

Figure 5. Shear Out Device for Coupler

Figure 6. Underframe following Static Crush test compared with Model

In order to validate the modelling predictions, each of the tests was modelled separately so that the model and experiment could be compared. The test components were modelled in a similar manner to the full underframe impact models, so that not only the testing models but also the full underframe models were validated by the experiments.

A fatigue test of the coupler was performed and finally full carriage impact testing was carried out and this successfully verified the designs against specification, see Figure 9.

Figure 7. Hillside Track for Impact Testing

Figure 8. Hammer Truck and Brake truck used in Impact Testing

Figure 9. Full Carriage Impact Testing

CONCLUSIONS

1. Mark 1 carriages have been identified as one of the most prone to overriding in a collision, and lend themselves to low-cost modifications to improve safety.

2. This paper details the modifications designed to improve the safety of Mark 1 vehicles. These modifications take the form of crush initiators, anti-climbing devices and shear-out components.

3. Each of the Mark 1 modifications has been tested under either static or impact conditions. The tests included static crushing of the underframe and impact testing of the anti-overriding device and the shear-out coupler.

4. The underframe crush testing results were comparable with the structural predictions giving confidence in both the design and the modelling.

5. The anti-overriding and shear-out coupler components have performed as expected under impact conditions. The cup and cone devices were seen to draw the headstocks together during impact and did not detach from the mountings until after this time. The shear-out couplers worked as desired during the high speed testing, and did not fail under 'rough shunt' loading.

6. Full carriage impact testing was carried out and this successfully verified the designs against specification.

REFERENCES

1. GM/RT2100 Railway Group Standard: Structural Requirements for Railway Vehicles. London. Railtrack Safety and Standards Directorate. Issue 2. July 1997.

S672/005/2000

Design and performance of energy absorbing subfloor structures in aerospace applications

A JOHNSON and **D KOHLGRÜBER**
German Aerospace Center (DLR), Institute of Structures and Design, Stuttgart, Germany

SYNOPSIS

Composite elements such as glass or carbon fibre reinforced tubes in axial crush exhibit high specific energy absorption and a challenge for the engineer is to utilise these properties in structural applications. The paper considers the application of composites in the design of helicopter and aircraft subfloors with a dual function of being structural elements under flight loads, which may be triggered to absorb energy by folding and crushing under crash loads. Examples discussed are an energy absorbing (EA) floor beam concept for a general aviation aircraft based on sandwich webs with glass and aramid fabric/epoxy face plates and box element core, and a modular helicopter subfloor concept with hybrid carbon/aramid/epoxy beams and cruciform elements. Progress on the development of design tools for these crashworthy structures based on FE simulation is also discussed and simulation results compared with crash test data.

1. INTRODUCTION

Composite materials are now being used in primary aircraft structures, particularly in helicopters, light aircraft, commuter planes and sailplanes, because of advantages such as light weight, high static and fatigue strength and their formability into large integral shell structures. However, the preferred materials such as carbon fibre/epoxy are inherently brittle and may exhibit a linear elastic response up to failure with little or no plasticity. By suitable choice of fibres, matrices and laminate layup, and by considering hybrid laminates with both brittle and ductile fibres, it is possible to develop composite structural elements which exhibit high mass specific energy absorption. The paper is concerned with utilising the EA properties of composites to design dual function aircraft structures, which are conventional structural elements under flight loads and are triggered to absorb energy under crash loads. In aircraft structures the main concern for the designer is to absorb crash energy associated with the vertical component of the impact velocity, since significant energy from the horizontal crash component may be absorbed by friction during aircraft sliding on soil or water surfaces. Emphasis is therefore placed on subfloor structures which absorb the vertical component of the impact energy and limit peak loads in the floor beams, seat rails and passenger seats. The paper considers examples of successful EA subfloor structures, and addresses in particular design tools available for predicting the crash response of these composite structural elements based on finite element (FE) simulation.

Before composites are used more extensively in aircraft structures it is important to understand the energy absorption and failure mechanisms, to have tools for assessing the crashworthiness of such structures, and to consider the design of special energy absorbing substructures or seats to maintain occupant safety levels. Conventional metallic structures absorb crash energy through plastic hinge formation and the folding or collapse of shell and beam elements and modern commercial FE codes, developed for automotive crash simulations, are able to simulate these effects. Two current EU funded research projects with aerospace industry participation are concerned with the development and validation of FE simulation tools to model the crash and impact response of composite structures [1], [2]. In these projects new materials models suitable for implementation into FE codes have been developed to characterise damage progression and failure modes in composites, and demonstrator subfloor structures have been fabricated and tested. In §2 the mechanical models and relevant test data for simulating structural response are discussed and energy absorbing mechanisms in composites are briefly described. Some recent examples of composite aircraft subfloor structures are then considered and their crash response compared with FE simulations. In §3 an energy absorbing EA sandwich beam concept for the aircraft keel beam of an executive jet, designed to reduce deceleration pulses on the seats, is analysed under impact loads. Section 4 is concerned with helicopter subfloor box structures fabricated from hybrid carbon/aramid fabric composites. Here the critical elements are cruciform nodes which connect the longitudinal and transverse floor beams. Comparison between predicted dynamic response and failure modes with observed test results are given in each case.

2. DETERMINATION OF COMPOSITE DESIGN PROPERTIES

2.1 Measured in-plane behaviour

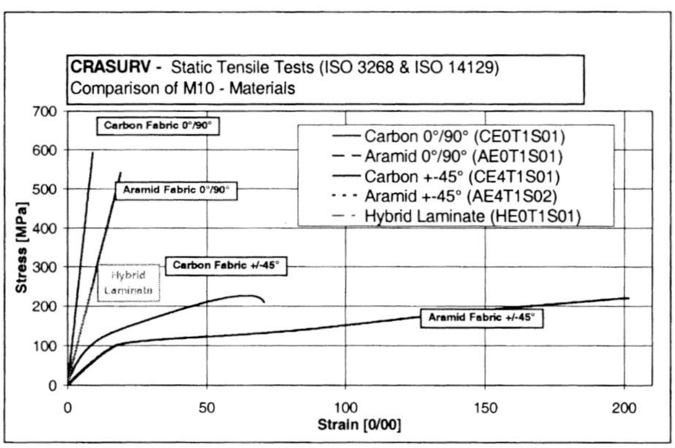

Fig 1: Tensile stress-strain curves for different carbon and aramid/epoxy laminates

The composites subfloor structures considered here are fabricated from carbon, glass and aramid fabric reinforced epoxy resins, and hybrid laminates containing carbon/aramid or glass/aramid plies. To obtain appropriate input data for the simulation of the structures and for the validation of the improved material models in FE codes, a comprehensive test program with both quasi-static and dynamic materials tests has been carried out within the CRASURV project and test data are summarised in [3]. Since the FE models are based extensively on laminated shell elements, the emphasis here is on in-plane ply properties for use in shell elements. Fig. 1 shows typical quasi-static tensile stress-strain curves for carbon and aramid fabric epoxy plies, and a hybrid laminate with layup $[A_{45}, A_{45}, C_{45}, C_0]_S$, where A and C refer to aramid and carbon plies, and the subscript is the fibre angle in the test specimen. The 45°-tension tests are used to characterise the ply in-plane shear response. The data show ply failure strains in the range 1-2% for carbon and aramid plies loaded in the fibre direction. The 45°-tension data are of interest for the design of EA structures since they indicate failure strains of over 20% in the aramid ply, with extensive nonlinearity and ductility. The tensile failure of hybrid specimens is controlled by the brittle failure of the central 0/90° carbon plies, although these hybrid laminates are much more ductile in bending and compression. These curves are used for the determination of parameters in the in-plane damage models discussed below.

2.2 Determination of composites EA properties

The high EA properties found in composite elements such as tubes in axial compression are not usually related to the above in-plane properties, since they are based on failure mechanisms such as delamination, fibre crushing, ply bending, etc. In order to compare EA levels in different composite systems, the DLR has developed a chamfered tube segment specimen, which is easy to fabricate and gives reproducible crush failures under quasi-static and dynamic axial loads. Fig. 2 shows a set of carbon, aramid and carbon/aramid hybrid segments before and after quasi-static axial crush tests. The dynamic tests were carried out in DLR's drop tower facility with initial velocities of 5m/s and 10m/s. The fabric/epoxy segments were tested with fibre orientations 0°/90° and ±45°, together with the hybrid lay-up. Fig. 3 shows typical load-deflection curves obtained in the test series. They have a characteristic shape with a steady crushing load after an initial load peak, indicating a high energy absorption.

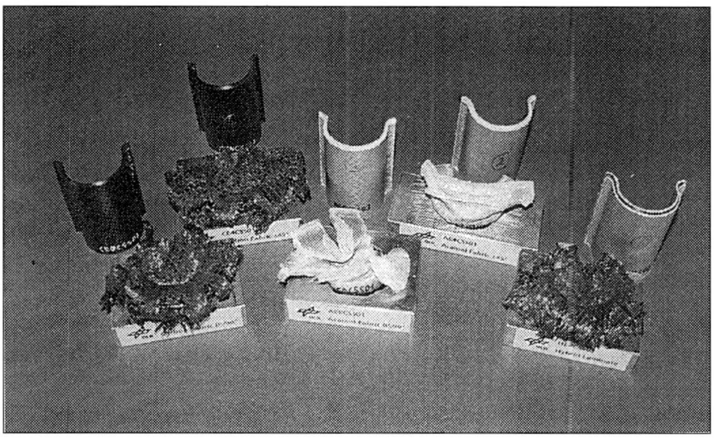

Fig 2: Crushing behaviour of different tube segment elements

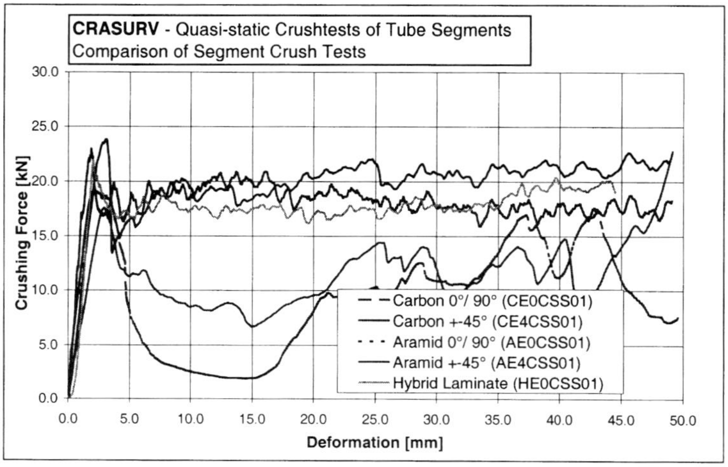

Fig 3: Comparison of measured load-deformation curves in segment crush tests

Study of the damaged segments shown in Fig. 2 and the load-deflection curves in Fig. 3 indicates that the carbon and aramid specimens exhibit different failure mechanisms and energy absorption levels. The carbon specimens absorb energy by a continuous crushing process after failure of the bevel trigger with almost constant crush load levels. The stiff carbon fibres are cracked into small debris, which disintegrate during the test. By contrast the aramid segments tend to behave like an elastic-plastic material, especially with the fibre orientation ±45° where a metal-like folding mechanism could be observed. These folding mechanisms always show higher periodic deviations in the load-deformation curves, leading to reduced energy absorption. In the hybrid segments energy absorbed is at the level of the carbon segments, but the tougher outer aramid plies provide some structural integrity and prevent the disintegration of the carbon debris. Although the fundamental failure or crushing mode was not influenced by an increase in test velocity, the measured loads decreased with increasing loading rate. In the dynamic tests on the carbon and hybrid segments the load level was reduced up to 25%. Table 2 summarises the average crushing stress and specific energy absorbed from the quasi-static and dynamic segment tests. Note the specific energies of 70 - 74.5 kJ/kg for the carbon and hybrid segments, which are at the upper end of the scale for composite tubes and well above similar data for aluminium and steel sections. The data confirm that the materials have the potential for successful applications in crash energy absorbing structures.

Table 1: Summary of results from quasi-static and dynamic segment crush tests

Segment Compression	Test Speed	CARBON G 803 / M10 0°/90°	±45°	ARAMID AT285 / M10 0°/90°	±45°	Hybrid Laminate
Average crushing stress σ_{avg} [MPa]	quasi-static	105.7	108.7	60.4	62.9	104.3
	v_0=5 m/s	81.9	82.5	52.5	67.1	83.0
	v_0=10 m/s	67.7	70.2	62.9	57.5	76.7
Specific absorbed energy E_{spm} [kJ/kg]	quasi-static	70.3	74.4	45.0	49.0	74.5
	v_0=5 m/s	55.6	55.0	39.7	50.8	57.5
	v_0=10 m/s	46.0	46.5	47.2	43.8	53.3
Crush force efficiency A_e [%]	quasi-static	80.2	82.4	51.6	60.6	81.2
	v_0=5 m/s	63.8	71.0	51.4	70.7	73.4
	v_0=10 m/s	71.8	78.7	72.3	70.1	77.0

2.3 Modelling composites properties

For metals there is extensive information in the literature on dynamic materials properties at large strains and high strain rates, and appropriate constitutive equations have been implemented into FE codes for structural impact simulations. For composite materials dynamic failure behaviour is very complex due to the different fibres and matrices available, the different fibre reinforcement types such as unidirectional (UD) fibres and fabrics, the possibility of both fibre dominated or matrix dominated failure modes, and the rate dependence of the polymer resin properties. Thus at present there are no universally accepted materials laws for crash and impact simulations with composites. It was considered that a homogeneous orthotropic elastic damaging material was an appropriate model for UD and fabric laminates, as this is applicable to brittle materials whose properties are degraded by microcracking. Constitutive laws for orthotropic elastic materials with internal damage parameters are described in [4], and take the general form

$$\varepsilon = S\,\sigma \qquad (1)$$

where σ and ε are vectors of stress and strain and S the elastic compliance matrix. In the plane stress case required here to characterise the properties of composite plies or shell elements with orthotropic symmetry axes (x_1, x_2), the in-plane stress and strain components are

$$\sigma = (\sigma_{11}, \sigma_{22}, \sigma_{12})^T \qquad \varepsilon = (\varepsilon_{11}, \varepsilon_{22}, 2\varepsilon_{12})^T. \qquad (2)$$

Using a strain equivalent damage mechanics formulation, the elastic compliance matrix S may then be written :

$$S = \begin{pmatrix} 1/E_1(1-d_1) & -\nu_{12}/E_1 & 0 \\ -\nu_{12}/E_1 & 1/E_2(1-d_2) & 0 \\ 0 & 0 & 1/G_{12}(1-d_{12}) \end{pmatrix} \qquad (3)$$

where ν_{12} is the principal Poisson's ratio, which for simplicity is assumed not to be degraded.

This general plane stress form for an orthotropic elastic material with damage has 3 scalar damage parameters d_1, d_2, d_{12} and 4 'undamaged ' elastic constants: the Young's moduli in the principal orthotropy directions E_1, E_2, the in-plane shear modulus G_{12}, and the principal Poisson's ratio ν_{12}. The damage parameters have values $0 \leq d_i \leq 1$ and represent modulus reductions under different loading conditions due to progressive damage in the material. Thus for unidirectional (UD) plies with fibres in the x_1 direction, d_1 is associated with damage or failure in the fibres, d_2 transverse to the fibres, and d_{12} with in-plane shear failure. For fabric reinforcements then d_2 is associated with the second fibre direction. In order to proceed further evolution equations are required which relate the damage parameters to other state variables. Several different models have been proposed in the literature. In [4], [5] it is postulated that the damage parameters are functions of strain energy release rates in the material, and this method has been developed further in the HICAS project [2], to include plasticity and rate effects in damage evolution.

This paper is concerned with the application of PAM-CRASH, a commercial explicit FE crash code [6], to analyse the crash response of composite structures. The code contains several materials models and special elements for laminated composite materials. The damage mechanics formulation outlined above is currently being implemented and tested in the HICAS project along with a new approach to delamination modelling described in [5]. The crash simulations of composite aircraft floor structures reported in §3 and §4 are based on alternative models in which it is supposed that damage evolution is dependent on the strain invariants, which can be determined from measured stress-strain curves such as those on Fig. 1. The elastic damaging materials law for fabric reinforcements may be modelled in PAM-CRASH as a 'degenerate bi-phase' model in which the UD fibre phase is omitted, and the 'matrix' phase is assumed to be orthotropic. If the simplifying assumption is made that $d_1 = d_2 = d_{12} = d$, the composite fabric ply has orthotropic stiffness properties, but a single 'isotropic' damage function d which degrades all the stiffness constants equally. The code does however allow different damage functions in tension and compression. This model has been found to be easy to apply and appropriate for quasi-isotropic laminates, which are commonly used in aircraft structures. An improved bi-phase model [3] has been developed and implemented in CRASURV, which allows two independent fibre directions in fabric plies and may be used when orthotropic damage effects are more significant.

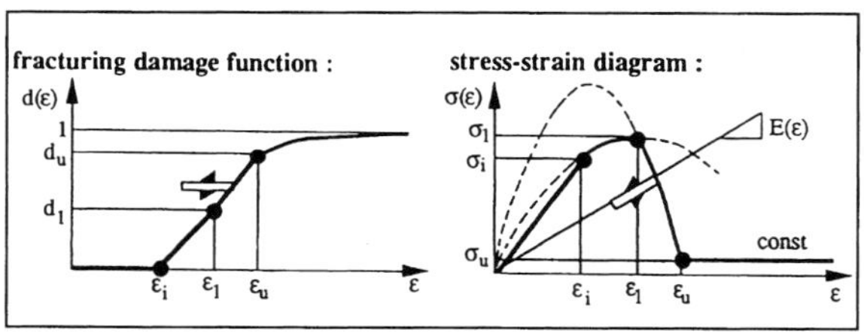

Fig. 4 Schematic fracturing damage function and corresponding stress-strain curve, [6].

It is necessary to determine parameters for the chosen composites model from measured stress-strain curves, such as Fig. 1. Uniaxial stress-strain curves are assumed to have a damage evolution equation which is a bilinear function of strain, in which there are two damage constants d_1 and d_u to be determined, (note d_1 here should not be confused with the damage parameter d_1 .) Typical uniaxial stress-strain curves have the general form shown in

Fig. 4, where ε_i is strain at the onset of initial damage, ε_1 is the strain at the peak failure stress, and ε_u is a limiting strain above which the stress is assumed to take a constant value σ_u. Measured test data for fabric composites are used to calibrate the materials model and to determine the damage parameters d_1 and d_u for the analysis. The parameter d_1 is related to the departure from linearity at the first 'knee' in the stress-strain curves, and is thus small in tension, whilst the parameter d_u determines the residual stress value σ_u. For the FE analysis it is not good practice to reduce the material stresses directly to zero at material fracture, as this may lead to numerical instabilities. Thus under tensile stresses typically $d_u \cong 0.95$, indicating that the element is nearly fully damaged, whilst in compression $d_u \cong 0.5$ to model the residual compression crushing stresses reported in Table 1. In this way it is possible to attribute measured crush stresses to shell elements under axial compression load so that energy absorption in composite structures may be simulated with shell models. Materials rate dependence is not included in the modelling presented here.

3. DYNAMIC CRUSH BEHAVIOUR OF SANDWICH FLOOR BEAMS

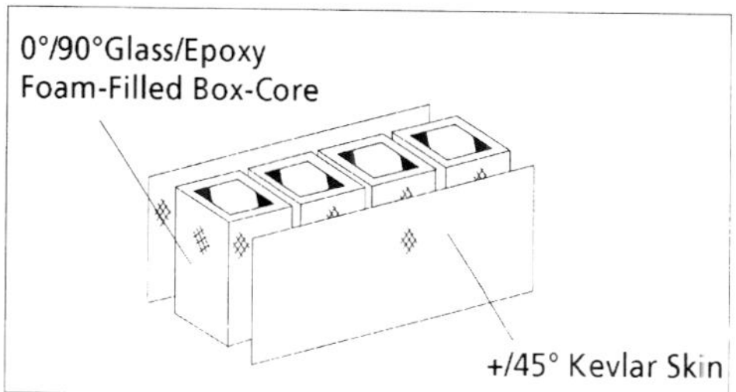

Fig. 5: Construction of box core beams

Crash tests carried out by NASA on general aviation (GA) aircraft with a carbon fibre/epoxy composite fuselage structure have shown that seat deceleration pulses are generally twice as high as those in a similar aluminium aircraft and too high for occupant safety. To meet current safety recommendations it is necessary to install safety seats and/or EA keel beams to support the seat rails. This led NASA to develop a modular EA floor beam concept for GA aircraft based on sandwich webs with glass and aramid fabric/epoxy faces and hollow GRP box cores or foam filled block cores, see Fig. 5. Kellas [7] describes in detail the beam design and fabrication concept, and has fabricated and impact tested box core and block core beams with 10 different cell geometries and skin layups, which demonstrated the success of this light weight EA beam concept. As Fig. 5 shows the beams themselves have a simple modular construction based on composite boxes, with or without foam cores, and by varying geometry and ply layup in the cells a wide range of floor beams with different stiffnesses and EA properties can be produced. One aspect of the beam design method concerned FE simulations with PAM-CRASH to determine failure modes and energy absorption in the beams under crash loads. Some typical results of the dynamic simulations are presented in this section on four selected beam constructions tested in [7].

In the nomenclature of [7] the beams are referred to as Box 3 & 4, and Block 3 & 4. All 4 beams have length 300 mm, height 200 mm, and thickness 25 mm, and consist of 5 cells with rectangular cross-section 60 mm × 25 mm. The block core beams are identical to the corresponding box core beam, but with the addition of a foamed PVC core. Box 3 and Block 3 contain glass fabric/epoxy (GRP) plies with fibres aligned at ± 45° to the box cell axes, whilst Box 4 and Block 4 have the glass fabric/epoxy with fibres at 0/90° to the box cell axes. The boxes are assembled as shown in Fig. 5, then the aramid fabric/epoxy (ARP) faceplates are added with fibres at ± 45° to the box cell axes. The laminate thickness depends on the beam fabrication method. For the GRP and ARP materials used a single ply has a nominal thickness of 0.25 mm. Reference to the beam construction shown in Fig. 5, indicates that the beam face sheets consist of 2 plies with GRP internal and ARP external, the end plates contain 2 GRP plies, and the internal webs contain 4 GRP plies.

The FE model of the box core beams contained about 5000 4-node layered shell elements to model the composite laminates. For the block core beams an additional 20000 8-node solid elements were added to model the foam core. It was assumed that the shell elements are attached to the outer faces of the solid elements, with common nodes. In the axial impact tests on the box core sandwich beams a rigid mass is attached to the top edge of the beam and they are accelerated downward onto a rigid base plate. In the PAM-CRASH model the beam was placed on a rigid wall and an impacting mass of 45 kg was distributed over the top edge nodes, which were additionally coupled in the axial direction and given an initial impact velocity. The impact velocities were chosen to be the same as in the tests and varied from 7.8 - 9.1 m/s. The test beams were supported laterally by two sets of knife edges, which were modelled as rigid plates. Since in impact the ARP plies are loaded principally at 45° to the fibre direction, the damage parameters were determined from tests on 45° specimens. When loaded in this direction ply properties are dominated by in-plane shear behaviour which leads to failure strains of about 5% in the GRP plies and 10% or higher in the ARP plies. These high failure strains in the ply materials in the main load direction help explain the failure modes observed in the beams, which were by folding of the side walls rather than the expected material fracture. Test data were also available on the PVC foams based on foam block compression tests. The foam was then modelled in PAM-CRASH as an elastic-plastic solid with a yield stress, and the composite shells with the default bi-phase model.

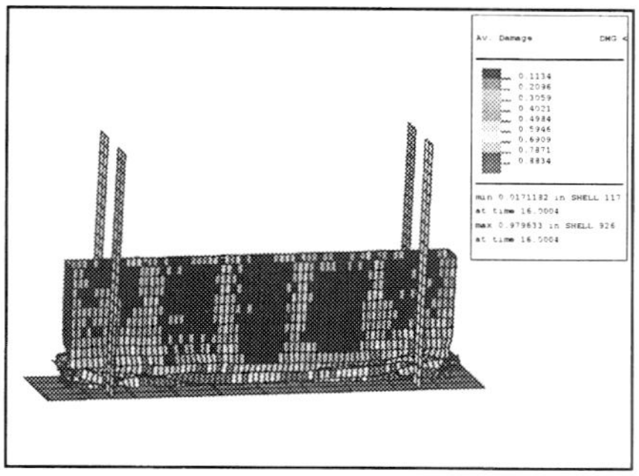

Fig. 6 FE simulation of box core beam

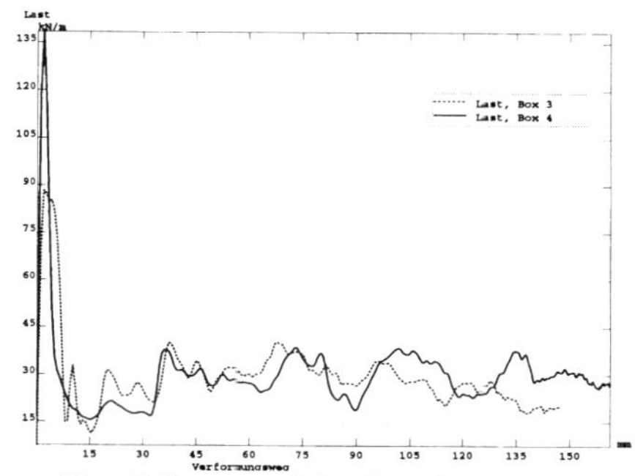

Fig. 7 Computed load-deflection response of Box 3 and Box 4 beams impacted at 7.8 m/s

The results of the PAM-CRASH simulations showed very satisfactory agreement with the beam tests reported. As the impact progressed the sidewalls collapsed in a folding mode as seen in Fig. 4 which shows the deformed Box 4 beam with the folds forming at the base, which is qualitatively the same as the observed failure modes in the impact tests, see [7, Fig. 11]. Fig. 7 shows the computed load-deformation response in the Box 3 and Box 4 beams. After a very high initial peak load, the dynamic crush load oscillates about at an approximately constant load level, which is a typical dynamic response of structures which fail by plastic hinge formation. The load-deflection curves have the same form as test data presented in [7]. A detailed comparison shows that the simulated initial peak is higher than the measured peak, however the mean loads and energy absorption predicted in the simulation are very similar to those measured, as can be seen in Table 2. The table lists the crush initiation loads (CIL), the steady crushing loads (SCL) and the energy absorbed (E_{ABS}) from tests and simulations for 4 beam structures. It is seen that the simulated peak CIL loads are well above the test loads, although the comparison here is critically dependant on the filtering used in both the test and simulation. More important is the generally good agreement between the crush loads and energy absorbed. The table shows in the final column the ratio of SCL from simulation and test, which are seen to be within 17 % of each other. Thus we conclude that the PAM-CRASH code is a valuable design analysis tool for understanding and optimising the properties of these modular floor beams.

Table 2: Comparison of loads and EA from tests and simulations

	Simulation with PAM-CRASH			NASA dynamic tests [6]			
Beam type	CIL kN/m	SCL kN/m	E_{ABS} kJ	CIL kN/m	SCL kN/m	E_{ABS} kJ	SCL ratio %
Box 3	83.31	26.96	1.31	67.55	27.23	1.40	99
Box 4	138.30	28.70	1.38	52.08	26.36	1.47	109
Block 3	108.00	35.67	1.36	90.82	42.96	1.81	83
Block 4	166.23	46.13	1.41	103.87	44.20	1.49	104

(CIL crush initiation load, SCL steady crush load, E_{ABS} energy absorbed)

3. CRASH RESPONSE OF HELICOPTER SUBFLOOR BEAMS

A helicopter design concept which meets the structural and crashworthiness requirements should provide a protective shell for the occupants, with energy absorbing elements incorporated in the landing gear, the subfloor and the seats. The subfloor typically consists of a framework of longitudinal beams and lateral bulkheads covered by the outer skin and cabin floor. The total structural height is often only about 200 mm. The design of intersections (cruciforms) of beams and bulkheads, the beam webs, outer skins, and floor sections (boxes) all contribute to the overall crash response of a helicopter subfloor assemblage. A generic helicopter subfloor box concept developed at the DLR within the CRASURV project is shown in Fig. 8, before and after a vertical crash test. A versatile FE model has been developed which allows different beam and cruciform elements to be used within a subfloor box, so that both the static structural integrity and the dynamic crush response of a range of different boxes may be simulated. In this section the dynamic crush response of some typical composite subfloor elements are simulated with PAM-CRASH as part of the design validation process.

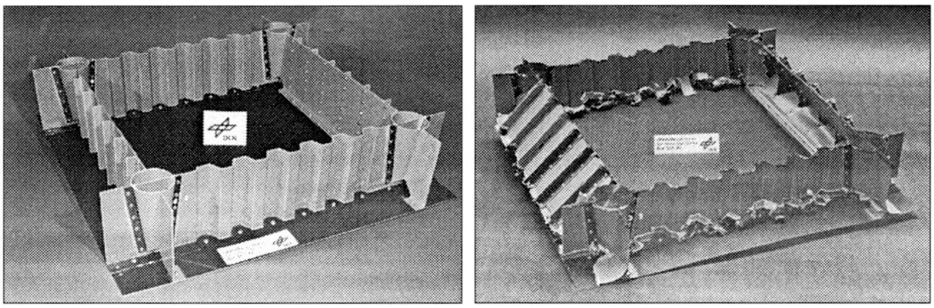

Fig 8: Hybrid sub-floor box before and after the dynamic drop test

A critical component in the sub-floor box is the cruciform corner element, since this is often a 'hard point' during crash loads which causes high loads to be transmitted through to the seat rails. The FE model of the cruciform element contained about 2000 4-node orthotropic layered shell elements to simulate the hybrid composite laminates, together with rigid body elements which simulated the rivets in the structure. A rivet failure model was not used, since in tests rivet failures never occurred. Structural tests on the cruciforms are carried out in a drop weight tower where the upper edges are embedded in an aluminium plate, which is impacted at about 10 m/s by a 50 kg mass. In the model the nodes of the upper edges form a rigid body with an added mass of 50 kg at the centre of gravity, and the base plate was modelled as a rigid wall. The cruciform element is designed to absorb energy under vertical impact loads, thus the composite materials selected are hybrid laminates of carbon and aramid fabric/epoxy. The laminate construction was based mainly on the symmetric hybrid 8 ply layup $[A_{45}/A_{45}/C_{45}/C_0]_S$ referred to in § 2.

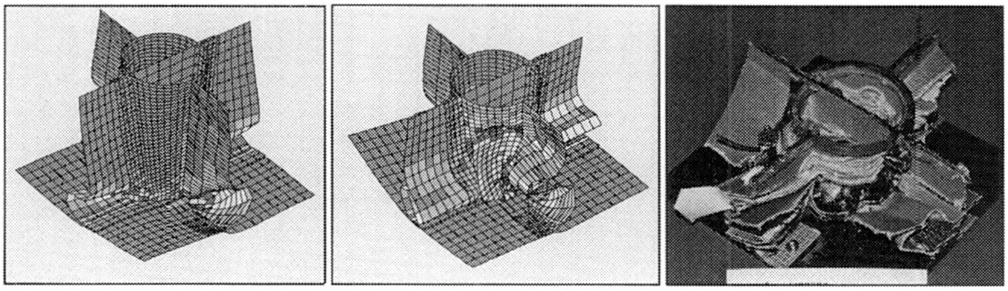

5 ms 15 ms

Fig. 9 Cruciform element: comparison of FE simulation with impacted element

The results of the dynamic simulations of vertical impacts on the cruciform elements showed excellent agreement between the predicted modes of failure and those observed in tests. In tests the element fails by regular folding of the vertical webs in contact with the base plate, initiated by the J-trigger at this position, which shows the success of this trigger concept. This is clearly seen in Fig. 9 in the FE simulation after 5 and 15 ms. Folding failures occur because of the hybrid laminate chosen, in which the more ductile aramid fabric plies are on the outer faces where the bending stresses are higher, which inhibits brittle failures associated with carbon fibre composites. A comparison of measured and simulated load-deflection response is given in [3] and shows good agreement up to 55 mm vertical displacement.

As a further example of the application of FE crash simulations in the sub-floor design procedure, the generic helicopter subfloor box in Fig. 8 was simulated under crash loads. The box consists of two trapezoidal corrugated longitudinal beams with embedded ply-drop-off triggers above the lower flange and two rib stiffened transverse beam sections. The transverse beam sections are connected by a conical split laminate to the longitudinal beams, forming cruciforms corner elements as in Fig. 9. The beam and cruciform elements are fabricated from hybrid carbon/aramid fabric laminates. In the drop tower impact test with a total mass of 501kg and a drop speed of 9.2 m/s, the trapezoidal beams and the central part of the intersections started to crush and absorb energy, the transverse beams fracture close to the trigger line and deflect outwards with low energy absorption, as is seen in Fig. 8.

A detailed FE-model of a quarter of this new box design was created with 8200 shell elements. The crushing behaviour was simulated with the improved bi-phase fabric model discussed in § 2, using parameters obtained from the materials test programme described there. The added mass was distributed along the nodes of the upper edges with impact velocity as in the test. The trapezoidal keel beams failed progressively in the region of the trigger, while the transverse beams tended to buckle outwards, in excellent agreement with the test structure. In Figure 10 the calculated load curve is compared to the load curve measured in the drop test. The integration of the contact forces between the skin of the sub-floor box and rigid wall located below the structure has been used for the determination of the predicted loads. The agreement with the mean crush load levels was reasonably good, but the pulse shape had poor agreement. It was considered that with these high impact masses there could by resonant vibrations in the drop tower load platform. Therefore, the FE model of the composite box was extended to include the steel drop tower load platform and the load cells situated at the platform corners, which were modelled using solid elements. In a final analysis with the load platform the load in the load cells was calculated and is also added to Fig. 10.

This is seen to correlate much better with the test results (curves 2 and 3), and shows the importance of simulating the complete dynamic system in this case and not just the test specimen.

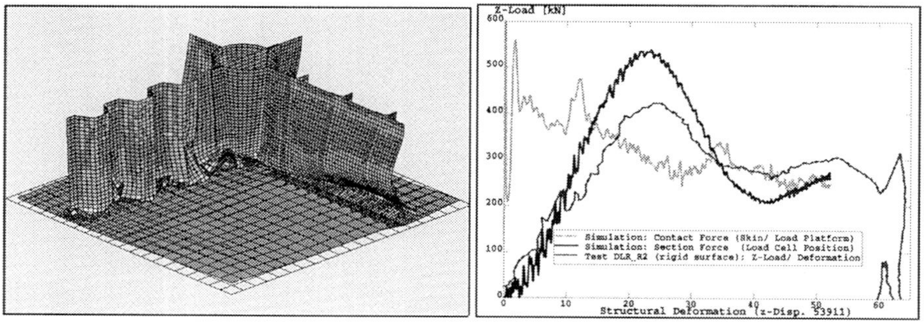

Fig. 10 Subfloor box: simulated deformation with predicted and measured load-deflections

5. CONCLUSIONS

The paper has demonstrated the application of prototype energy absorbing composite elements in GA aircraft and generic helicopter subfloors. Further work in CRASURV has led to the development of larger composite subfloor structures based on sine-wave beam concepts for commuter and transport aircraft. However, the conventional aircraft industry development process based on design, fabricate, test and certify is too time consuming and expensive for such innovative structures, since the variety of composite materials, fabrication techniques and structural concepts is extensive. Thus simulation tools for predicting the crash response of composite structures are required first for screening design concepts, and then for reducing prototype crash testing and certification programmes. Simulation methods for predicting crash and impact response of composite structures are being developed within two EU projects supported by the aircraft industry [1], [2]. Materials models for composite shell elements with fibre fabric reinforcement, which are implemented in the explicit FE code PAM-CRASH [6] were summarised in the paper. Simulation of composite subfloor beam structures under low velocity crash loading conditions were performed to validate the code and materials models. Good agreement was obtained with crush loads and energy absorption for glass/aramid/epoxy sandwich beams. In a more complex helicopter subfloor structure with carbon/aramid hybrid laminates, failure modes, predicted loads and energy absorption were well modelled. The FE simulation methods are generally good at predicting failure modes in composite structures, but further refinement is needed for better correlations with measured peak loads and energy absorbed. In ongoing work improvements to composites models for crash and impact simulation are being developed, with particular attention directed to delamination models, plasticity effects in shear and materials rate dependence.

ACKNOWLEDGEMENTS
Part of the work presented here was developed in the EU projects CRASURV [1] and HICAS [2]. The authors wish to acknowledge the financial contribution of the CEC, and the project partners for their scientific contributions.

REFERENCES

[1] CRASURV Design for Crash Survivability, EU RTD Project BE 95-2033, 1996-2000.
[2] HICAS High Velocity Impact of Composite Aircraft Structures, EU RTD Project BE 96-4238, 1998-2000.
[3] D. Kohlgrüber, *Numerical simulation of composite aircraft structures under crash loading,* CEAS Forum on Crash Questions, CIRA, Capua, Italy, Feb. 14-15, 2000.
[4] P. Ladevèze, E. Le Dantec, *Damage modelling of the elementary ply for laminated composites*, Composites Science and Technology, **43**, 257-267, 1992.
[5] A.F. Johnson, A.K. Pickett, *Impact and crash modelling of composite structures: A challenge for damage mechanics,* European Conf. Computational Mechanics, ECCM'99, Munich, 1999.
[6] PAM-CRASH™ FE Code, Engineering Systems International, 20 Rue Saarinen, Silic 270, 94578 Rungis-Cedex, France.
[7] S. Kellas, *An experimental investigation into the EA performance of composite beam webs for aircraft subfloor applications,* American helicopter Society NTS Meeting on Rotorcraft Structures, Williamsburg, 1995.

S672/007/2000

Energy absorption in hybrid composite structures

G REYES VILLANUEVA and **W J CANTWELL**
Materials Science and Engineering, University of Liverpool, UK

ABSTRACT

The quasi-static and impact properties of a novel fibre-metal laminate based on a tough glass fibre reinforced polypropylene (GFPP) matrix have been evaluated and compared with those of a conventional thermosetting based carbon fibre reinforced epoxy (CFRP). Initially, the mode I, mode II and mixed-mode I/II interlaminar fracture properties of the plain composites were investigated. Here, it was shown that the mixed-mode interlaminar fracture toughness of the GFPP system was close to 5000 J/m^2 and that of the thermosetting system was approximately 2000 J/m^2. Single cantilever beam tests indicated that the adhesion in the GFPP/aluminium and the CFRP/aluminium systems was excellent at all loading rates and that failure frequently occurred within the composite rather than at the bi-material interface.

Low velocity impact tests on both types of fibre metal laminate highlighted the superior impact response and energy-absorbing characteristics of the thermoplastic-based system. Here, it was shown that the GFPP system absorbed significant amounts of energy through matrix cracking in the composite plies and membrane stretching in the aluminium layers. It is believed that thermoplastic-based fibre metal laminates offer significant potential for use in energy-absorbing engineering structures.

INTRODUCTION

Fibre-metal laminates systems consisting of thin layers of high strength aluminium alloy bonded to plies of fibre-reinforced epoxy resins are presently attracting the interest of a number of aircraft companies. Fibre-metal laminates such as ARALL, GLARE and CALL, combine the isotropic properties, ease of fabrication and durability associated with metals with the high strength and fracture resistance of lightweight composite materials. Previously published work has shown that fibre-metal laminates offer an excellent resistance to fatigue loading [1], superior impact properties [2] and excellent damage-tolerance characteristics [3].

Young *et al* [4] have shown that fibre-metal laminates offer substantial improvements in fatigue life and residual strength improvements of up to forty percent compared to their traditional metallic counterparts.

Vlot [5] conducted a number of impact tests on a range of fibre-metal laminates and showed that for a given impact energy, the size of the damage zone in these multi-layered laminates was considerably smaller than those observed in the plain composite materials. Lawcock *et al* [6] showed that laminates with lower levels of fibre surface treatment exhibited increased amounts of fibre/matrix splitting and offered superior post-impact residual tensile properties.

Clearly, the processing cycle required to achieve complete consolidation of thermosetting-based FML's can be very long and cost-ineffective. One means to overcome these limitations is to use a thermoplastic-based composite that can be bonded to the metallic layers and shaped in a simple one-shot stamping operation. This procedure will clearly shorten the processing cycle resulting in significant reductions in manufacturing costs.

The aim of the present work is to investigate the mechanical response of a glass fibre reinforced polypropylene/aluminium fibre-metal laminate. Here, attention will focus on characterising the quasi-static and impact properties of this novel lightweight material. The potential offered by these novel systems will be further investigated through a series of comparative tests on a conventional thermosetting-based fibre metal laminate.

EXPERIMENTAL PROCEDURE

The fibre-metal laminates investigated in this research programme were based on a 0.5 mm thick glass fibre reinforced thermoplastic (Plytron from Borealis Ltd, Norway) and a 0.4 mm thick woven carbon fibre reinforced rubber-toughened epoxy (Stesapreg EP121-C15-53 from Stesalit Ltd, Switzerland). Initially, the mode I, mode II and mixed-mode I/II interlaminar fracture energies of the composite materials were determined using the double cantilever beam (DCB), the end notch flexure (ENF) and the mixed-mode flexure (MMF) geometries respectively. The carbon fibre/epoxy (CFRP) laminates were prepared by stacking 16 plies of composite in a picture frame mould and heating the stack in a hot press for 90 minutes at 120°C. The GFPP laminates were manufactured by placing 10 plies of unidirectional tape in a mould, heating to 185°C in an air-circulating oven before stamping in a cold press. The subsequent rapid cooling process ensured a low degree of crystallinity in the polypropylene matrix [7]. Prior to laminating, a starter defect based on a folded aluminium starter defect was incorporated at the mid-plane of both laminates. In preparation for testing, 20 mm wide samples were removed from the plates using a diamond slitting wheel and their edges were ground to remove any scratches or other types of damage.

The interlaminar fracture tests were conducted on an Instron 4505 at a crosshead displacement rate of 1 mm/minute. Crack propagation in the DCB and MMF geometries was monitored with the aid of a calibrated scale painted on the edge of the specimen. No attempt was made to monitor crack propagation in the ENF samples. The mode I fracture energy, G_{Ic}, was determined using Berry's compliance method for which:

$$G_{Ic} = \frac{nP\delta}{2Ba}$$

where n is determined from the slope of the plot of log compliance vs log crack length, P is the applied force, δ is the displacement, a the crack length, and B the specimen width.

The mode II interlaminar fracture energy was determined using:

$$G_{IIc} = \frac{9a^2P\delta}{2B[2L^3 + 3a^3]}$$

where L is the half span of the beam. The mixed-mode fracture energy was calculated using:

$$G_{I/IIc} = \frac{21P^2a^2C}{2B[7a^3 + 2L^3]}$$

The fibre-metal laminates were manufactured by stacking the appropriate number of composite and aluminium plies in a picture frame mould. Optimum adhesion between the glass fibre/polypropylene composite and the aluminium layers was ensured by applying an amorphous chromate coating to the as-received aluminium alloy. In addition, a 0.26 mm thick interlayer based on a maleic anhydride modified polypropylene (Fusabond M613-05 from DuPont de Nemours) was incorporated at the composite-metal interface during the manufacturing operation. Prior to laminating the carbon fibre/epoxy system, the aluminium alloy was subjected to a simple abrasion/wipe procedure in which the alloy was abraded with a 1200 grit paper and cleaned with acetone. The fibre-metal laminates were then manufactured by following the aforementioned manufacturing cycle for the appropriate composite.

Initial testing on the fibre-metal laminates focused on characterising the fracture properties of the bond between the aluminium alloy and the composite. Here, single cantilever beam (SCB) tests were conducted using the bi-material configuration shown in Figure 1. The SCB samples were manufactured by stacking either 5 GFPP or 8 CFRP plies on a two millimetre thick aluminium sheet. A folded aluminium foil was incorporated at the composite-metal interface to act as a starter defect. SCB specimens with dimensions of 20 x 200 mm were removed from the moulded plates and clamped in a steel fixture. Tests were conducted at crosshead displacement rates between 1 mm/minute and 3 m/s. At low loading rates, crack extension was monitored optically with the aid of a graduated scale on the edge of the sample. At intermediate and high rates, the initial and final specimen compliance was measured and an inverse compliance procedure was used to determine the crack length for a given crosshead displacement. The interfacial fracture energy was determined using a compliance calibration procedure where the specimen compliance, C, was determined at various crack lengths, a, and a curve fit of the following form was applied:

$$C = C_o + ka^3$$

where 'C_o' and 'k' are constants for a particular specimen. Here, 'k' was determined by measuring the slope of the graph of the compliance versus the cube of the crack length. The interfacial fracture energy, G_c, was then determined using:

$$G_c = \frac{P^2}{2B}\frac{dC}{da}$$

Subsequent testing was conducted on the multi-layer laminates shown in Table 1. Here, a 2/1 laminate refers to a structure consisting of two 0.8 mm thick aluminium layers either side of a composite ply. Similarly, a 3/2 laminate consists of 3 layers of aluminium and two layers of composite stacked alternatively. In preparation for tensile testing, dog-bone samples with a 10 mm wide working section were removed from the plates. Tensile tests were conducted on an Instron 4505 universal test machine at a crosshead displacement rate of 1 mm/min. The strain in the samples was measured using a clip-on extensometer.

The fracture characteristics of the fibre-metal laminates were evaluated through a series of single edge notch bend (SENB) tests on specimens with dimensions 80 x 15 mm x thickness. Prior to testing, a 7.5 mm long pre-crack was introduced into at mid-span. The specimens were positioned on two simple supports positioned 60 mm apart and loaded at a crosshead displacement rate of 1 mm/minute.

The energy absorbing characteristics of the FML's were investigated using a falling weight impact tower. Here, a 2 kg carriage was released from heights of up to 0.8 metres on to plates supported on circular ring with an internal diameter of 75 mm. A piezoelectric load cell located just above the 12.7 mm diameter impact head was used to measure the impact force during each test. The residual tensile strength of the damaged laminates was subsequently evaluated by removing samples with dimensions 200 x 50 mm from the plates and conducting tension tests at a crosshead displacement rate of 1 mm/minute. Finally, a number of samples were sectioned, polished and examined under a low magnification optical microscope.

RESULTS AND DISCUSSION

Interlaminar Fracture Tests on the Plain Composites

The interlaminar fracture properties of both composites were evaluated by conducting mode I, mode II and mixed-mode I/II tests at a crosshead displacement rate of 1 mm/minute. Figure 2 summarises the results of these tests. It is interesting to note that the mode I interlaminar fracture toughnesses of both systems are relatively high, with the GFPP system offering an average value of approximately 2500 J/m^2 and the CFRP material a value of 1900 J/m^2. An examination of the edges of the samples highlighted the presence of extensive fibre bridging in the polypropylene composite whereas the woven nature of the fibres in the CFRP material suppressed this phenomenon. Similarly, the mode II interlaminar fracture toughnesses of both systems were similar with the thermoplastic-based composite exhibiting a value of G_{IIc} of approximately 3800 J/m^2 and its thermosetting counterpart a value of 4100 J/m^2. Significant differences were observed in the mixed-mode interlaminar fracture properties of the two materials with the thermoplastic significantly out-performing the thermoset (4900 J/m^2 compared to 2150 J/m^2). A detailed SEM investigation of the fracture surfaces of the failed MMF samples highlighted the presence of significant plastic flow in the polypropylene composite whereas only limited regions of localised ductility were observed in the carbon fibre reinforced epoxy. Nevertheless, these preliminary interlaminar fracture tests suggest that both of these systems offer an excellent interlaminar fracture toughness and both should be ideal candidates for incorporation into fibre metal laminates.

Single Cantilever Beam Tests
The fracture properties of the interface between the composites and the aluminium alloy were evaluated using the SCB geometry shown in Figure 1. Figure 3 shows typical quasi-static load-displacement curves for SCB samples based on both types of material. An optical examination of the polypropylene-based composite indicated that the crack propagated in a stable manner along or close to the composite-metal interface involving the generation of significant amounts of fibre bridging. Crack propagation in the CFRP samples also occurred in a stable manner although small instabilities in which the crack jumped several millimetres were observed in most specimens. Here, little or no fibre bridging was observed in the wake of the primary crack. After testing, many of the load-displacement curves exhibited a small residual displacement as a result of plastic deformation within the aluminium substrate.

The load-displacement curves in Figure 3 were then used to determine the interfacial fracture energy of each specimen and typical resistance curves at 1.0 mm/minute are shown in Figure 4. From the figure, it is clear that the fracture energy for the GFPP/aluminium system rises gently with increasing crack length before plateauing at a value close to 3500 J/m^2. This value of interfacial fracture toughness is clearly very high and comfortably exceeds the interlaminar fracture toughness of tough thermoplastic-based composites such as carbon fibre reinforced PEEK [8]. It is believed that the rising R-curve was associated with the development of the fibre bridging mechanism reported above. In contrast, the CFRP-based system did not exhibit any pronounced R-curve and the interfacial fracture energy remained roughly constant at a value of approximately 2000 J/m^2.

Figure 5 shows the variation of the plateau values of the interfacial fracture toughness with crosshead displacement rate for both composite systems. At the lowest rate of loading, the interfacial fracture toughness of the GFPP-based FML was 3550 J/m^2. With increasing loading rate, the fracture energy increases quite sharply to an average value of 4700 J/m^2 at 100 mm/minute. At high rates, the interfacial fracture energy begins to fall, reaching a value of approximately 3000 J/m^2 at 3 m/s. These trends in the interfacial fracture energy suggest that there may be competing failure mechanisms occurring during fracture in this material. This effect was investigated further by examining the fracture surfaces of the failed samples. Figure 6 shows low power optical micrographs of the aluminium substrate following tests on the GFPP system at loading rates between 1 mm/min and 3 m/s. At 1 mm/minute, Figure 6a, it is clear that in certain regions, the crack has propagated largely within the fibre-reinforced composite. From the figure, it is clear that a number of fibres are slightly distorted, probably as a result of localised flow during the moulding process. At 100 mm/minute, Figure 6b, the crack remains largely within the composite although small regions of exposed aluminium were in evidence. Here again, there is evidence of appreciable amounts of fibre movement during the manufacturing operation. Under impact conditions, the fracture surface again exhibits a mixed appearance with the crack oscillating between the interface and the composite, Figure 6c. This evidence suggests that the slight decrease in the fracture energy of the GFPP-based material at high crosshead displacement rates may be associated with a greater tendency for the crack to propagate at or close to the composite-aluminium interface.

Figure 5 also includes the variation of the interfacial fracture toughness of the CFRP/aluminium samples with crosshead displacement rate. The data show that for low and intermediate loading rates, the interfacial fracture energy is relatively high for an epoxy-based system, typically 2000 J/m^2. This value is close to the mixed-mode value of $G_{I/IIc}$ measured on the plain composite suggesting that failure in these bi-material samples is closely linked to this parameter. Although the value of G_c under impact conditions was approximately twenty-five percent lower than the quasi-static value, the level of adhesion is still encouraging under this dynamic loading condition. Figure 7 shows low magnification optical micrographs of the aluminium substrates of SCB specimens tested at 1 mm/min., 100 mm/min. and 3 m/s. An examination of the fracture surface of the specimens tested at 1 and 100 mm/minute indicates that residual composite is present on the surface of the aluminium suggesting that the crack has propagated almost entirely within the tough composite material. A similar observation was made at 3 m/s although small regions of aluminium are apparent across the fracture surface. This evidence suggests that a very strong interface between the composite and the aluminium alloy can be achieved by applying a very simple abrasion/wipe preparation to the surface of the aluminium prior to lamination.

Tensile Properties of Fibre-Metal Laminates

The tensile elastic modulus and failure stress of the fibre-metal laminates was measured on standard dog-bone samples at a crosshead displacement rate of 1 mm/minute. Figure 8 shows typical stress-strain curves for the three GFPP fibre metal laminates. Also included in the figure are typical stress-strain curves for the plain aluminium and composite. All of the FML stress-strain curves exhibited a linear response up to the point at which the composite layers fractured and the load dropped dramatically. Further loading of the these FML's resulted in the progressive failure of the aluminium plies until complete fracture of the samples occurred at relatively high strains. A subsequent optical examination of the 2/1 and 3/2 samples showed that the composite plies had fractured along a path perpendicular to the applied stress initiating planes of delamination at the composite-aluminium interface. Localised fibre fracture was also observed in the 4/3 laminates although very little delamination was in evidence long the bi-material interfaces. The stress-strain curves for the carbon fibre/epoxy fibre metal laminates were similar in appearance to those shown in Figure 8. An optical examination of the CFRP-based specimens indicated that failure in all laminates initiated in the composite layers. In the 2/1 and 3/2 laminates, the primary crack propagated across the composite to the bimaterial interfaces before propagating along the length of the sample in a slow and stable manner. In order to complete the tensile data, a limited number of tests were conducted on a 2/17 laminate. In these samples, failure occurred in a catastrophic manner involving brittle failure in both the composite and aluminium plies without inducing any delamination at the composite-metal interface.

Figure 9 shows the variation of the tensile modulus of the FML's with volume fraction of composite. Also included in the figure are the elastic moduli of the plain aluminium alloy and the plain composites. The data clearly indicate that there is a roughly linear relationship between the modulus of the laminates and the volume fraction of the composite with the modulus of both systems tending to decrease with increasing V_f. The figure clearly demonstrates that, as expected, the carbon-based laminates offer a superior modulus for a given volume fraction of composite. The solid lines in Figure 9 correspond to a law of mixtures prediction based on the elastic moduli of the aluminium and composite constituents. Clearly, agreement between the experimental results and the law of mixtures prediction is excellent for both sets of material across the range of systems examined. The evidence suggests that the tensile modulus properties of fibre metal laminates similar to those considered here can be predicted accurately from a knowledge of the tensile properties of the appropriate constituents.

The variation of the tensile strength of the fibre metal laminates with volume fraction of composite is shown in Figure 10. Here, incorporating greater amounts of composite into the FML has a positive effect with the strength increasing rapidly with increasing V_f. For the GFPP system, the tensile strength of the 4/3 laminate with a volume fraction of 43% is approximately 70% greater than that of the plain aluminium alloy. Here again, the tensile properties of the laminates exhibit low levels of scatter and appear to follow a simple law of mixtures rule. From the figure, it is clear that the tensile strength of the CFRP system is superior to that of the GFPP system and that a rule of mixtures approach would prove to be inaccurate in this case. For this material, increasing the volume fraction of the composite to almost fifty percent resulted in a 120% increase in the tensile strength.

Single Edge Notch bend Tests

Figure 11 shows the variation of the work of fracture with volume fraction of composite for the two FML systems. The figure shows that incorporating glass fibre reinforced PP into the aluminium serves to increase the work of fracture quite significantly. An examination of the fractured specimens highlighted the presence of extensive fibre pull-out and splitting in the vicinity of the primary crack path. These tests were repeated at impact rates of loading where an even steeper rise in the work of fracture with increasing volume fraction was observed. In contrast, increasing the amount of carbon fibre reinforced epoxy in the FML's resulted in a decrease in the work of fracture. This reduction is due in part to the fact that fifty percent of the fibres in the woven fabric were aligned in the direction of crack propagation and did not therefore contribute in any significant manner to the fracture process

The Impact Behaviour of Fibre-Metal Laminates

In this part of the study, low velocity impact tests were conducted on the 2/1, 3/2 and 4/3 laminates at energies up to 7.5, 10 and 20 Joules respectively. An examination of the cross-section of the 2/1 GFPP-based system, Figure 12a, indicates that the plate exhibits significant permanent deformation around the point of impact. A closer examination of this region highlights the presence of a large number of matrix cracks pointing away from the point of impact. A detailed study indicated that these matrix cracks extend through regions with higher concentrations of fibres possibly following paths along the fibre-matrix interface. Further inspection indicated that a plane of delamination was present between the top surface aluminium layer and the composite ply whereas the lowermost composite-aluminium interface remains intact. The 4/3 laminates exhibits a conical-shaped fracture zone similar to that observed in many plain composites, Figure 12b [9]. Again the evidence in this micrograph suggests that the interface between the composite and metal plies maintains its outstanding toughness characteristics at very high rates of loading.

Figure 13 shows micrographs of the cross-section of two CFRP fibre metal laminates. The micrograph of the 2/1 laminate, Figure 13a, shows that a small area of damage has initiated directly under the point of impact. Here, a plane of delamination and large transverse crack propagating through the composite layer are apparent. In addition the micrograph shows that the aluminium plies have undergone localised plastic deformation, a mechanism that is likely to absorb considerable energy. The micrograph of the 4/3 laminate, Figure 13b, highlights similar failure mechanisms although a large plane of delamination is clearly visible between the uppermost aluminium and composite plies. The planar damage areas in all of the damaged laminates were measured as a function of the incident impact energy. Figure 14a shows the variation of the damage area with impact energy for the 3/2 laminates. The figure indicates that the carbon-based system offers a superior impact resistance at low energies whereas the glass fibre/PP system offers superior impact characteristics in the higher energy range. Figure 14b shows the variation of damage area with impact energy for the 4/3 laminate. Here, the superiority of the thermoplastic-based system is clearly evident with the GFPP FML exhibiting a damage of approximately 500 mm^2 after a 20 Joule impact.

Figure 15 summarises the residual tensile properties of the impact-damaged 4/3 FMLs. The data clearly indicate that the glass fibre/polypropylene laminate offers superior residual properties at all impact energies. However, it is interesting to note that the properties of the CFRP FML closely mirror those of its thermoplastic counterpart. An optical examination of laminates subjected to low and intermediate impact energies showed that final fracture occurred at locations away from the impact zone, often near the zone where the specimens were gripped. This evidence suggests that thermoplastic-based fibre-metal laminates exhibit an excellent resistance to localised impact loading and offer great potential for use in dynamically-loaded structures. One further advantage of these systems is the likelihood that low and moderate amounts of impact damage can be easily repaired using fusion bonding techniques similar to those used to join thermoplastic components. Previous work has shown that impact-damaged thermoplastic-matrix composites such as carbon fibre reinforced PEEK can be repaired by applying heat and pressure to the damaged region [10]. It is believed that similar techniques can be applied to repair the type of thermoplastic FML investigated in this study.

CONCLUSIONS

The mechanical properties of a novel fibre-metal laminate based on glass fibre reinforced polypropylene composite have been compared with those of a traditional carbon fibre reinforced epoxy FML. Single cantilever beam tests on both systems have shown that excellent adhesion between the composite and metal plies can be achieved if the appropriate pre-treatments are employed. Tensile tests on a number of stacking configurations have shown that the carbon fibre FML offers superior tensile strength and modulus properties at quasi-static rates of strain. However, falling weight impact tests on a simple plate configuration have shown that the thermoplastic-based FML offers an excellent resistance to low velocity impact loading. A detailed study of the impacted samples indicated that the thermoplastic laminate absorbed energy in localised matrix cracking and membrane stretching in the aluminium plies whereas as the epoxy-based system absorbed energy in fibre fracture, delamination and membrane stretching.

ACKNOWLEDGEMENTS

The authors are grateful to the Mexican Government (CONACYT) for supporting this work. The donations of the glass fibre reinforced polypropylene by Sandra Coburn of BI

Composites, the carbon fibre reinforced plastics by Stesalit Ltd. and the Fusabond M613-05 by Dr. Karlheinz Hausmann, DuPont de Nemours, Switzerland are also gratefully acknowledged.

REFERENCES

[1] Gunnick J.W. and Van der Shee P.A., Comp. Struct. 4, Ed. I.H. Marshall, Elsevier App. Sci., 1987. pp1.162–1.177.
[2] Vlot A. and Fredell R.S., Proc ICCM 9, Ed. A. Miravete, Woodhead Publ., 1993, pp51–58.
[3] Macheret J. and Bucci R.J., Eng., Frac. Mech. 45, 1993, pp729–739.
[4] Young J.B., Landry J.G.N., and Cavoulacos V.N., Composite Structures 27, 1994, pp457–469.
[5] Vlot A., Int. J. of Impact. Engineering, 18, 1996, pp291–307.
[6] Lawcock G.D., Ye L., Mai Y.W., and Sun C.T., Comp. Sci. and Tech. 57, 1997, pp1621–1628.
[7] Davies P. and Cantwell W.J., Composites 25, 1994, pp869–877.
[8] Davies P., Cantwell W.J., Jar P.-Y., Richard H., Neville D.J., and Kausch, H.H. ASTM STP 1110 Composite Materials: Fatigue and Fracture, Ed. T.K. O'Brien, pp70–88, 1991.
[9] Boll D.J., Bascom W.D., Weidner J.C., and Murri W.J., J. Mats. Sci. 21, 1986, pp2667–2677.
[10] Cantwell W.J., Davies P., Jar P.-Y., Bourban P.E., and Kausch H.H., Proc. 11th European SAMPE Conf. Basel 1990, Ed. H.L. Hornfeld, pp411–426.

Table 1. Summary of the laminates studied in this research programme.

Composite	Laminate	Configuration	Thickness (mm)
CFRP	2/1	2 layers of Al + 1 composite ply	1.9
CFRP	3/2	3 layers of Al + 2 composite plies	3.0
CFRP	4/3	4 layers of Al + 3 composite plies	4.1
GFPP	2/1	2 layers of Al + 1 composite ply	2.5
GFPP	3/2	3 layers of Al + 2 composite plies	4.3
GFPP	4/3	4 layers of Al + 3 composite plies	6.0

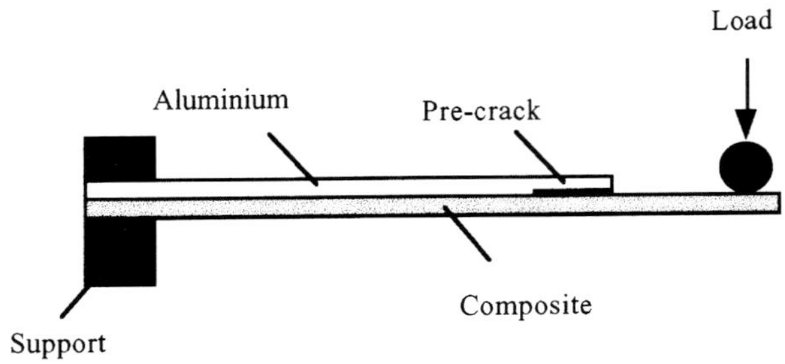

Figure 1. Schematic of the single cantilever beam geometry.

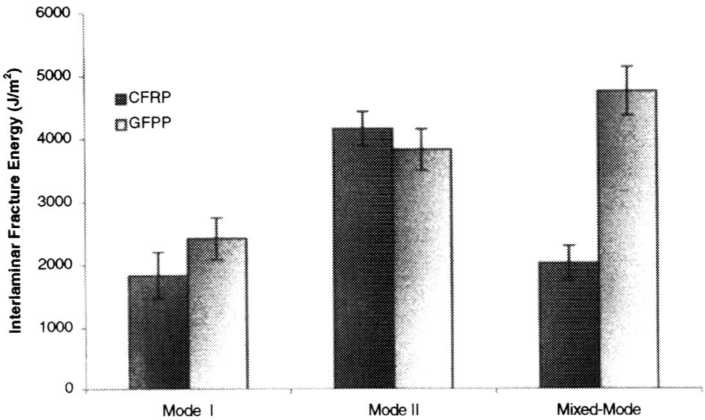

Figure 2. Summary of the mode I, mode II and mixed-mode interlaminar fracture properties of the plain glass fibre/PP and carbon fibre/epoxy composites.

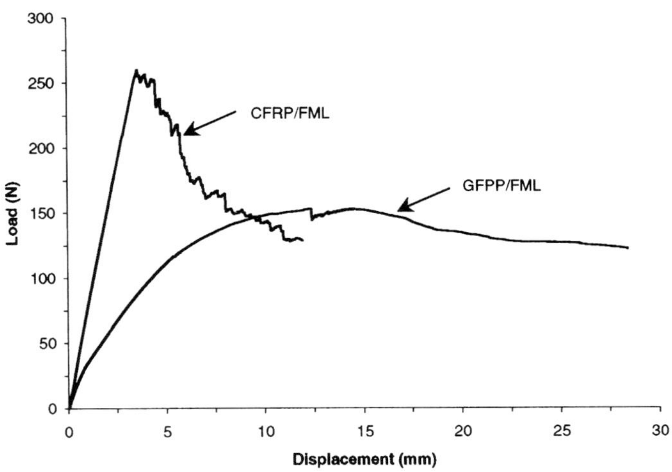

Figure 3. Typical load-displacement curves following single cantilever beam tests on CFRP/aluminium and GFPP/aluminium samples.

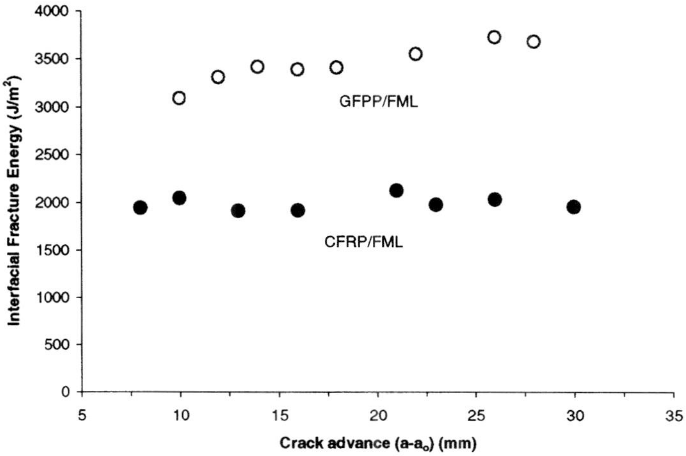

Figure 4. Typical resistance curves following single cantilever beam tests on CFRP/aluminium and GFPP/aluminium samples.

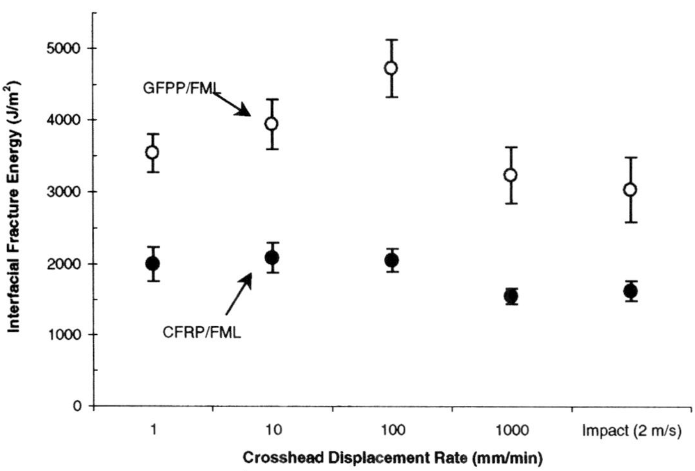

Figure 5. The variation of the interfacial fracture toughness with crosshead displacement rate for both types of FML.

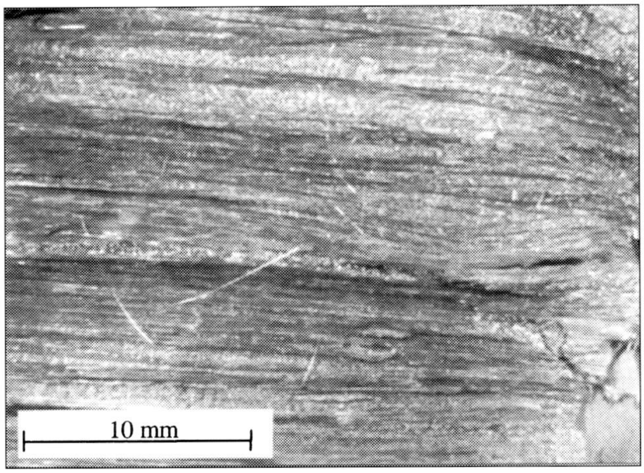

(a) **1 mm/min**

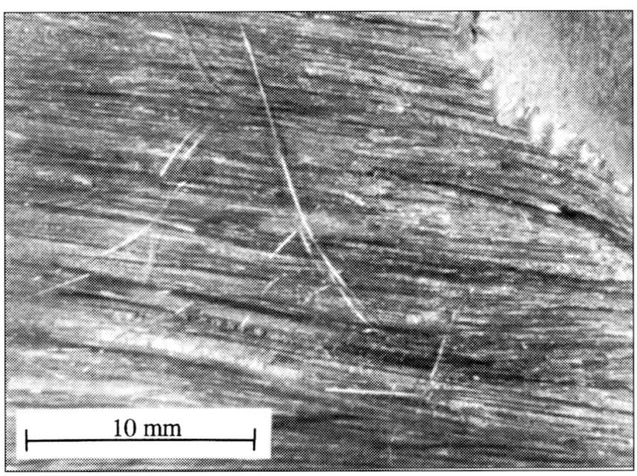

(b) **100 mm/min**

Figure 6. Low magnification optical micrographs of failed glass fibre/PP SCB specimens.

(c) **Impact 3 m/s**

Figure 6. (continued) Low magnification optical micrographs of failed glass fibre/PP SCB specimens.

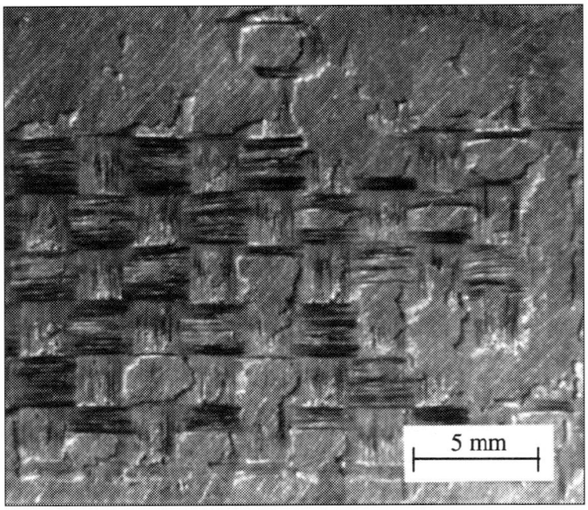

(a) **1 mm/min**

(b) **100 mm/min**

Figure 7. Low magnification optical micrographs of failed carbon fibre/epoxy SCB specimens.

(c) **Impact (3 m/s)**

Figure 7. (continued) Low magnification optical micrographs of failed carbon fibre/epoxy SCB specimens.

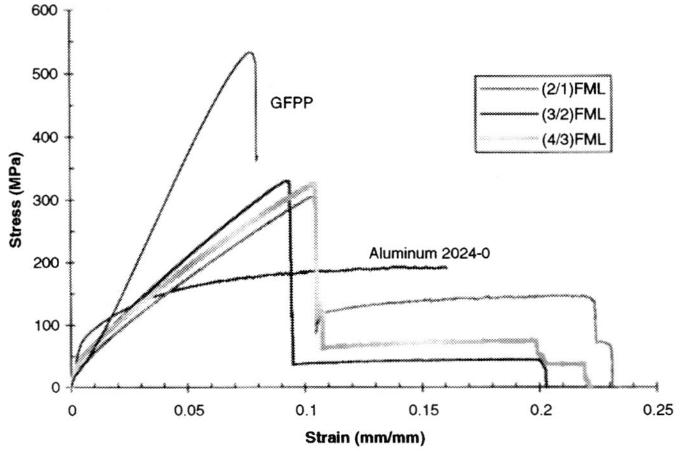

Figure 8. Typical tensile stress-strain curves for three glass fibre/PP fibre-metal laminates.

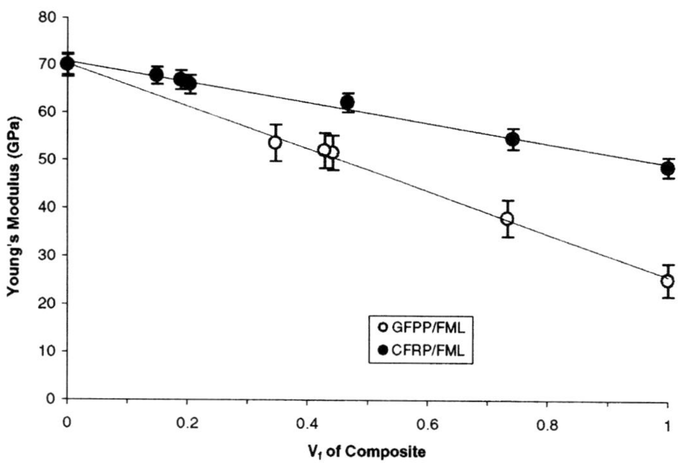

Figure 9. The variation of the tensile modulus of the fibre-metal laminates with volume fraction of composite.

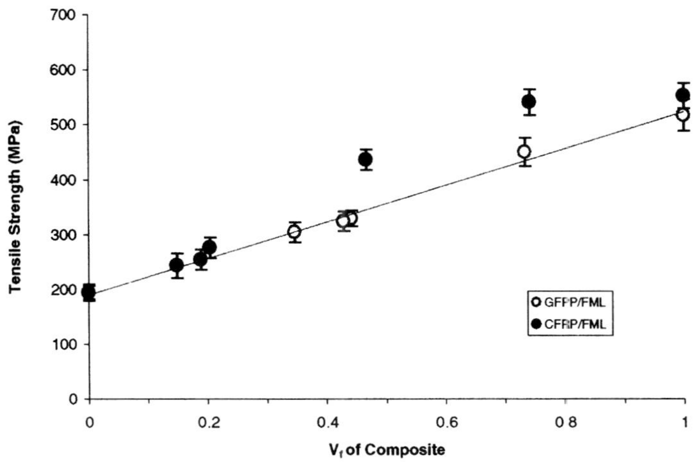

Figure 10. The variation of the tensile strength of the fibre-metal laminates with volume fraction of composite.

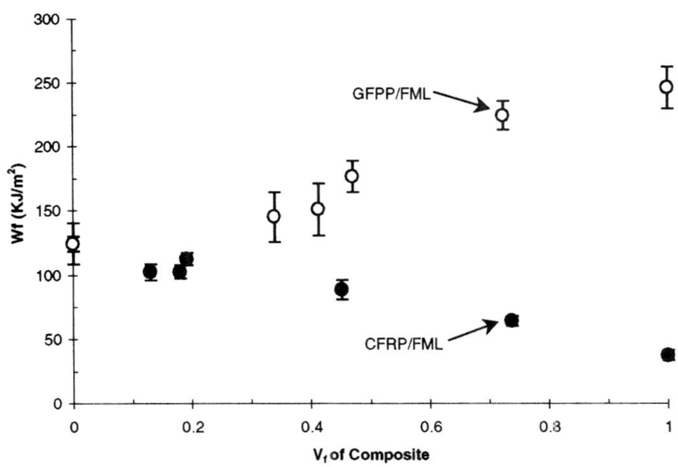

Figure 11. The variation of the work of fracture with volume fraction of composite for both families of FML.

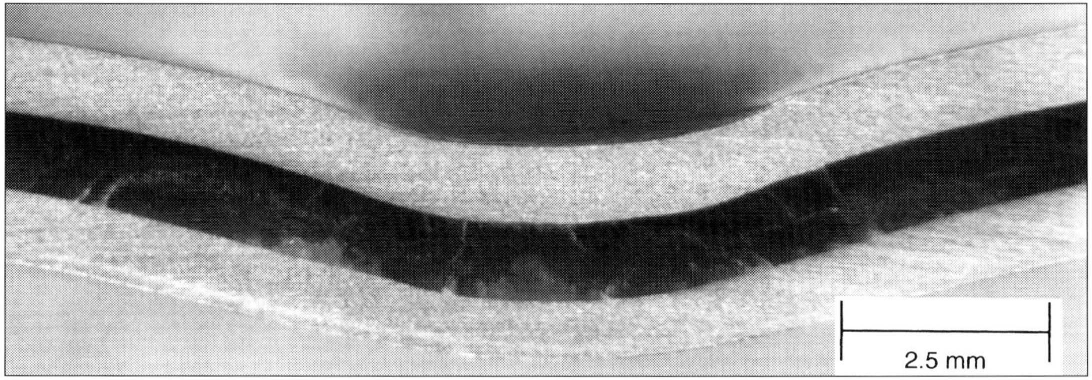

(a)

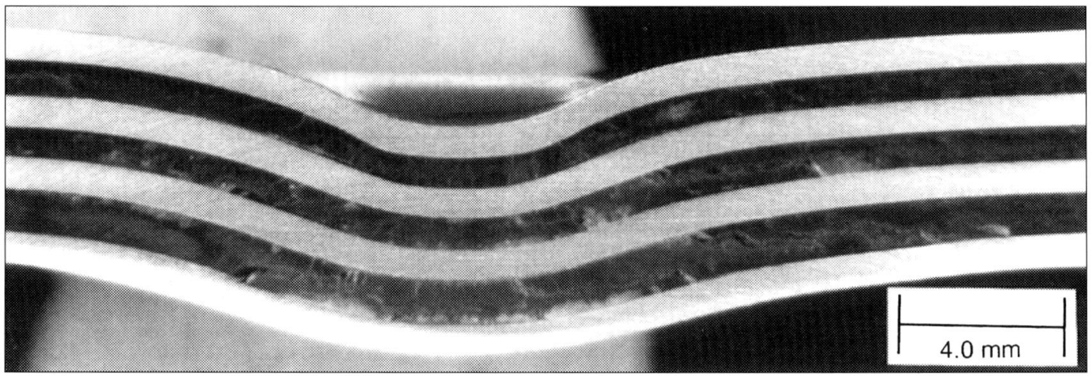

(b)

Figure 12. Low power optical micrographs showing damage in a GFPP laminate
(a) a 2/1 laminate subjected to an impact energy of 7.5 Joules
(b) a 4/3 laminate subjected to an impact energy of 20 Joules.

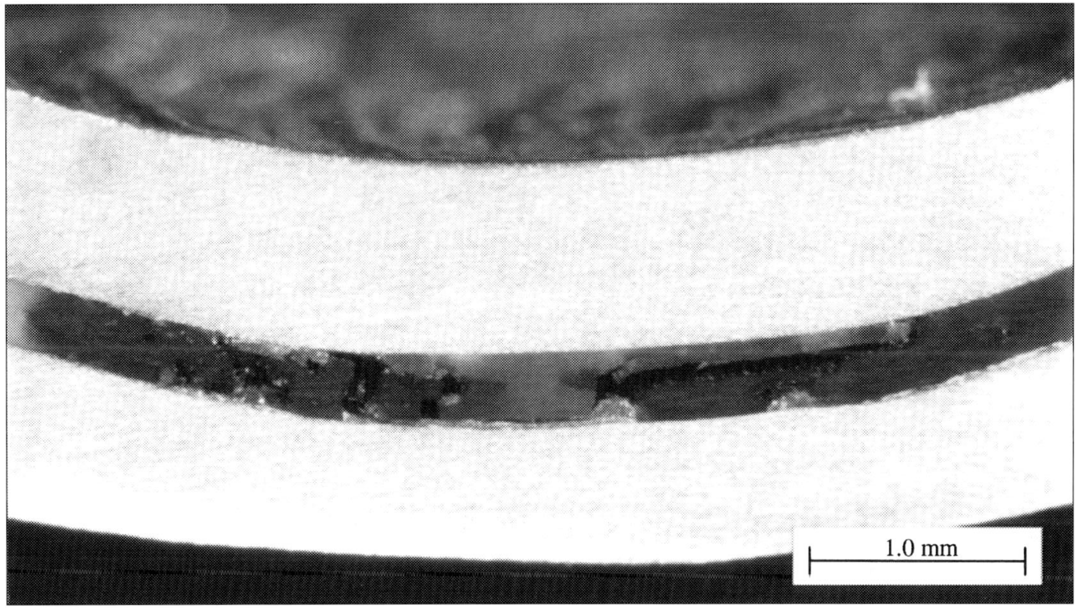

(a) **2/1 Laminate (Impact energy = 7.5 Joules)**

**Figure 13. Low power optical micrographs showing damage in a CFRP laminate
(a) a 2/1 laminate subjected to an impact energy of 7.5 Joules**

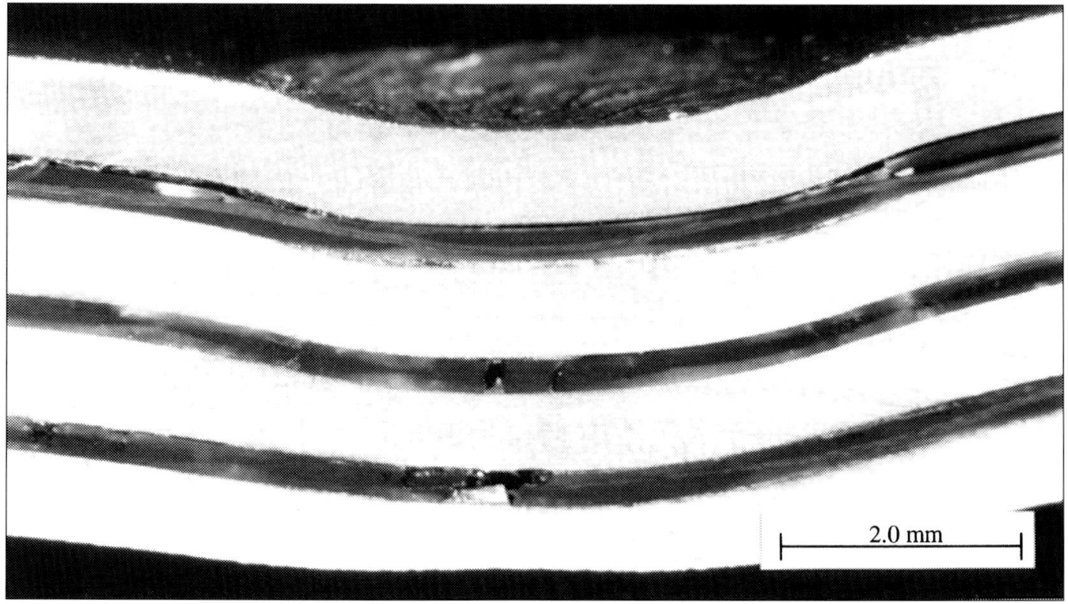

(b) **4/3 Laminate (Impact energy = 20 Joules)**

Figure 13.(continued) Low power optical micrographs showing damage in a CFRP laminate
(b) a 4/3 laminate subjected to an impact energy of 20 Joules.

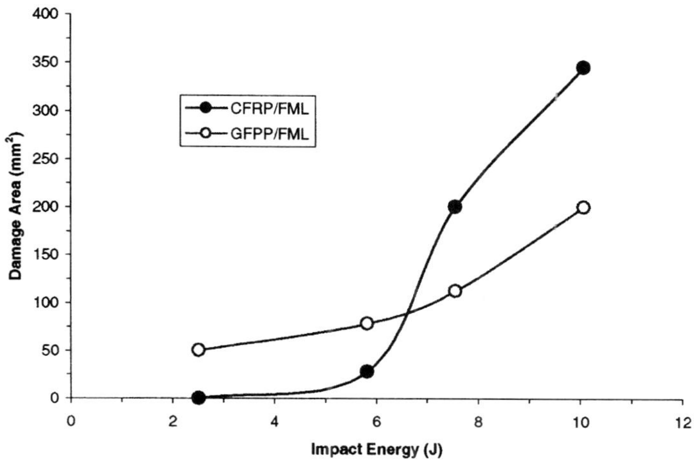

Figure 14. The variation of damage area with impact energy for the GFPP and CFRP fibre-metal laminates (a) 3/2 laminates

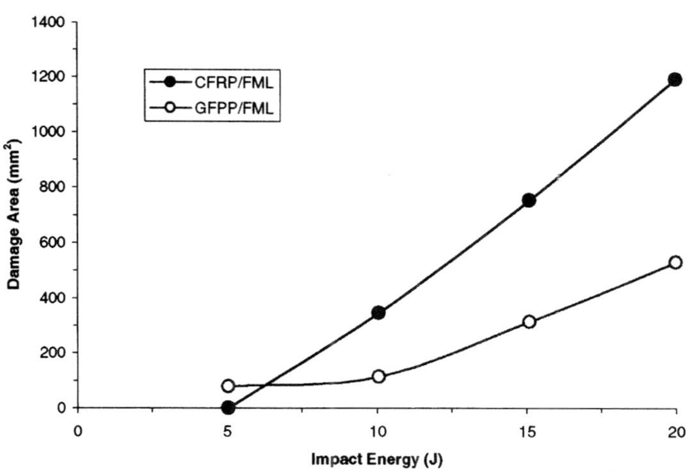

Figure 14. (continued) The variation of damage area with impact energy for the GFPP and CFRP fibre-metal laminates (b) 4/3 laminates.

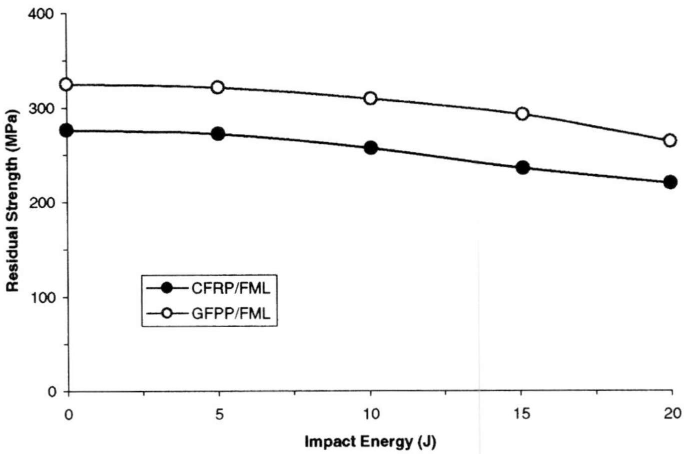

Figure 15. The variation of the residual tensile strength with impact energy for the 4/3 laminates.

S672/008/2000

Development of autobody sheet materials for crash performance

M G MARSH
Corus Group plc, Crewe, UK

SYNOPSIS

Increasingly stringent crash legislation and the drive to produce lightweight car bodies has promoted the extensive use and development of high strength steels. In this paper a number of material properties which are thought to contribute significantly to energy absorption in metallic materials will be highlighted. It will be shown that it is important to understand the relationship between strength and thickness in order to downgauge confidently. The increased dynamic stability exhibited by high strength steels will also be demonstrated. Finally, a brief example will be shown of how the collapse resistance can be increased through the simple manipulation of the joining process. The conclusion from the work is that to maximise energy absorption within a component, it is important to take into account the interaction of the material, the section design, the forming process and the final fabrication technique.

1. INTRODUCTION

To meet the challenge of reduced vehicle mass with no compromise in performance, the steel industry has developed, and is continuing to develop, high strength steel products. The application of high strength steels has proven to be highly successful with their extensive introduction into many current automotive vehicles and also those which are under development. However, to meet the increasing demands placed on vehicle safety, a more thorough understanding of energy absorption, in terms of both the material and its application is required.

A wide range high strength steel sheet products are now commonly available, many developed over the last decade. For crashworthiness, multiphase materials such as dual phase and TRIP steels have received particular interest, as they exhibit increased energy absorption compared to other steels of equivalent strength. The ability of a metallic material to absorb energy is a function of a number of properties which include its static strength, work hardening and response to strain rate. The understanding of the influence of the component design is also of importance as it influences the strains, strain history and strain rates developed during an impact event. In addition, the actual deformation witnessed in practice can be extremely complicated and yet material selection continues to be based on high speed uniaxial tensile properties.

The material thickness is an important parameter which influences the collapse resistance of a structural member, and if the automotive designer is to downgauge confidently then the interaction of steel strength and thickness needs to be fully understood. In addition, the reduction in sheet thickness may also influence the collapse stability and component stiffness. The former in some respects is a function of the design and the materials utilised whilst the latter may be influenced by adopting continuous joining techniques, tailor welded blanks or hydroforming. These latter techniques have been clearly demonstrated to provide extensive lightweighting potential in conjunction with high strength steels [1]. This paper will briefly highlight and discuss the issues introduced above with reference to a few examples from available literature and ongoing research programmes.

2. STRAIN RATE SENSITIVITY AND ENERGY ABSORPTION

Although high strength steels provide an opportunity for effective gauge reduction, it is generally found that as the steel strength is increased the strain rate sensitivity reduces. This is clearly illustrated in Figure 1, where the dynamic factor, defined, in this instance, as the ratio of the material dynamic strength (at 200 s^{-1} strain rate) to the static strength, is plotted as a function of the nominal yield and tensile strengths of several steel materials.

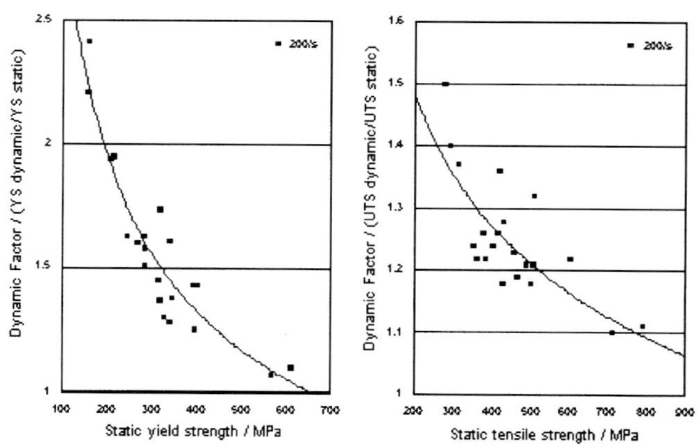

Figure 1. Plot displaying the change in the dynamic factor as a function of static strength.

In both instances the dynamic factor, which is a direct measure of the materials strain rate sensitivity, is shown to reduce with increasing steel strength. This is more readily shown in

Figure 2, where the change in the yield and tensile strength with strain rate is displayed for two different materials, the first a typical mild steel and the second a high strength dual phase material.

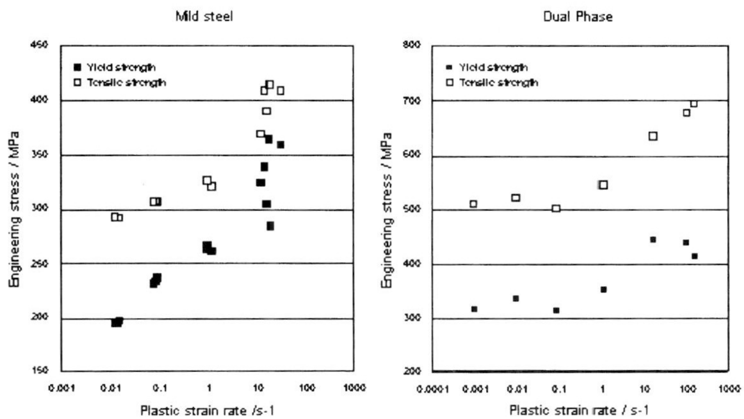

Figure 2 Change in strength as a function of strain rate.

It can be observed that the mild steel is significantly more sensitive to strain rate than the higher strength dual phase product. The yield strength of the mild steel is also shown to be more sensitive than its tensile strength which is in contrast to the behaviour of the dual phase where the tensile strength displays a slightly increased response relative to its yield strength. This latter observation is in agreement with other authors [2] who found that the n-value of high strength dual phase steels tended to increase with increasing strain rate. Subsequently, the dual phase material is able to retain and even increase its work hardening response under dynamic conditions, hence facilitating energy absorption.

Due to these observed differences in the response of the two steels it is unsurprising that different strain rate laws suit different steels. The different behaviour also indicates the possible reasons for some contradictions within the literature [2,3,4,5]. Some authors measure the sensitivity as a change in the yield strength whilst others use the change in tensile strength. Yet further work has used the change in the energy absorbed within the dynamic tensile test, i.e. the area under the stress/strain curve. This representation of material response is most likely a more suitable measure of performance, as the energy absorbed is a function of the static strength of the material, the strain rate sensitivity and the work hardening exhibited. The energy absorbed by a number of different steel grades in uniaxial tensile tests, under both static and dynamic conditions (200 s^{-1} strain rate) is shown in Figure 3.

Although, high strength steels have reduced strain rate sensitivity it is clearly apparent that they are able to absorb significant amounts of energy. However, this response changes with different strain rates. As a consequence it is therefore important to identify the relevant level of strain rates within automotive structures. This will in turn allow optimum material selection decisions to be made. Recent work carried out by Corus suggests that for an impact test (18 ms^{-1}) of a simple section, the strain rates occurring above approximately 200 s^{-1}

contributed very little to the overall energy absorbed. Although, high strain rates were witnessed they were highly localised and therefore of little influence. This is an important finding as some authors carry out dynamic tensile testing at strain rates as high as 2000 s^{-1} whilst others limit material assessment at strain rates upto 200 s^{-1}. The significance of this is that a different choice of material may be made depending on the level of strain rate used for comparison.

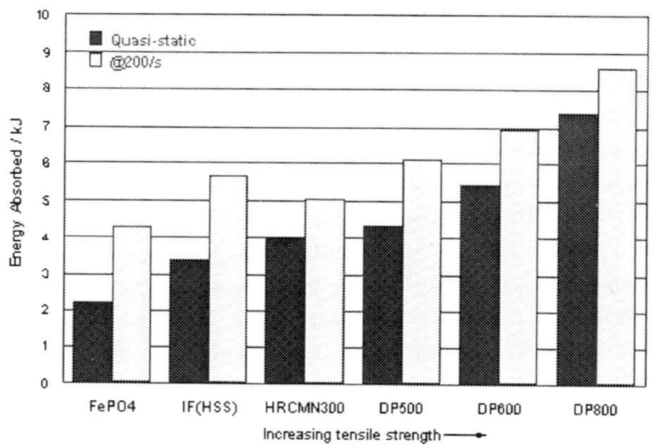

Figure 3. The area under the stress/strain curve upto 10% elongation for various steel grades.

However, there are possibly other material properties which are equally important for energy absorption that are not readily identified by high speed uniaxial tensile testing. Previous work carried out by the author [6] found that the material within a collapsing structure was subjected to a bending and unbending type of deformation with significant amounts of compressive strain. The strain reversal properties of the material, and whether it displays significant kinematic or isotropic hardening, may therefore also be important in dictating the average collapse force under dynamic conditions. Due to this observation the current test methods upon which materials are selected for crash energy management need to be reviewed. The design of the component plays an important part in dictating the type of loading the material witnesses. Finite element analysis would facilitate such selection by the identification of strains, their magnitude and direction, strain histories and strain rates in an iterative process.

Although the strain rate sensitivity of high strength steels is reduced compared to mild steel products, its characterisation within the material model used in crash analysis is critically important for accuracy. This is clearly illustrated in Figure 4.

A simple rectangular hydroformed component was experimentally tested under both impact and quasi-static conditions. Dynamic analysis of the section was also carried out with and without strain rate sensitivity. It is shown that the model which includes strain rate sensitivity predicts the dynamic experimental test more closely with the correct prediction of the observed folding modes and therefore collapse performance.

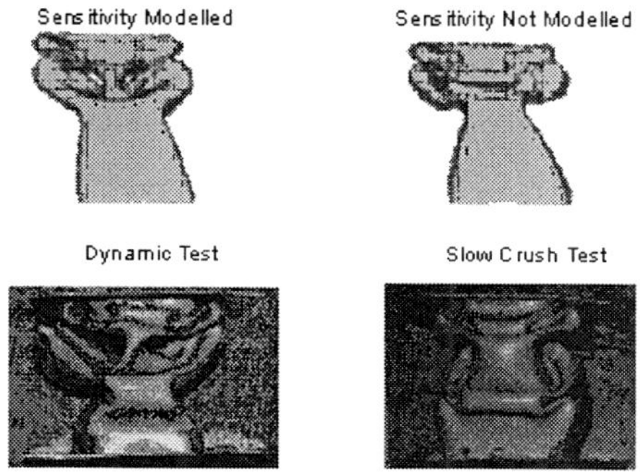

Figure 4. Utilisation of strain rate sensitivity in FE analysis and its influence on the predicted mode of collapse, after ref. [6].

The discussion above provides an indication of the properties which are necessary for effective energy absorption and the methodology in making informed decisions with regard to appropriate materials. A steel which would be an effective energy absorber would have a high static strength, a positive response of the tensile strength to the strain rate, (upto approximately 200 s^{-1}), and significant dynamic work hardening, in addition to isotropic hardening properties. In order to facilitate proper material selection the identification of the relevant strains and strain rates occurring in the component is necessary. This can be achieved through FE modelling which requires the proper definition of the materials strain rate properties. This makes optimum material selection an iterative design process.

However, the above discussion concentrates on the nominal properties of the material. In reality, the final properties within the component can be very different depending on the manufacturing process. The influence of the forming operation can be witnessed even on simple structures, as shown in Figure 5.

In the figure the peak and average collapse loads measured during the slow crush of a simple mild steel, 60 mm square section, 1.2 mm in thickness manufactured by sheet metal bending and hydroforming are compared. The peak load is shown to be significantly influenced by the forming operation. The accurate prediction of the peak load is an important requirement in automotive design. In this simple example the strains introduced by the hydroforming process have significantly altered the response of the structure. The response of the material to the forming operation therefore needs careful consideration, and may possibly even be exploited to provide further enhancement in the overall performance.

3. SHEET THICKNESS

In order to downgauge confidently to facilitate lightweighting, an understanding of the relationship between steel strength and sheet thickness is required. Sperle and Lundh [7] evaluated the impact performance of a range of steel strengths and sheet thickness. They suggested that the following expression can be used to define the influence of tensile strength and steel gauge on the energy absorbed by simple square sections;

$$E = 0.158\ Rm^{0.506}\ t^{1.498}$$

where E = absorbed energy for 150 mm deformation
t = thickness
Rm = tensile strength

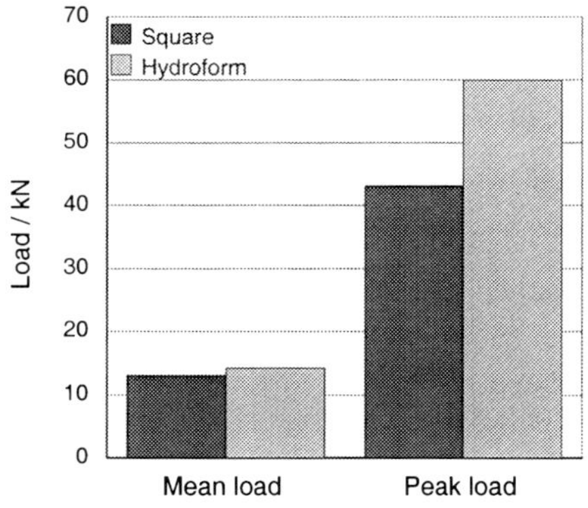

Figure 5. The collapse performance of simple square sections

Recent work by the author, however, indicates that the response to thickness actually changes with different steel grades. Using a simple closed top hat structure a range of steels with varying thicknesses have been examined. The resulting relationship for the mean collapse load at 30 mph was found to be;

$$Pav = k\ Rm^{0.4}\ t^{a}$$

where Pav = mean collapse load
k = constant related to the geometry
a = a range of values between 1.6 and 2.0

This change in the response to thickness with different steel grades needs to be further evaluated with a wider range of materials in order to identify the most effective grades for downgauging.

In both of the above relationships the influence of thickness is found to be significantly more influential than steel strength. Substituting the tensile strength in the above equation by a term which better describes the working strength of the material in collapse, i.e. the flow strength, and using $t^{1.8}$, the mass saving potential as a function of increasing material strength for a given collapse load, may be determined. This is shown in Figure 6.

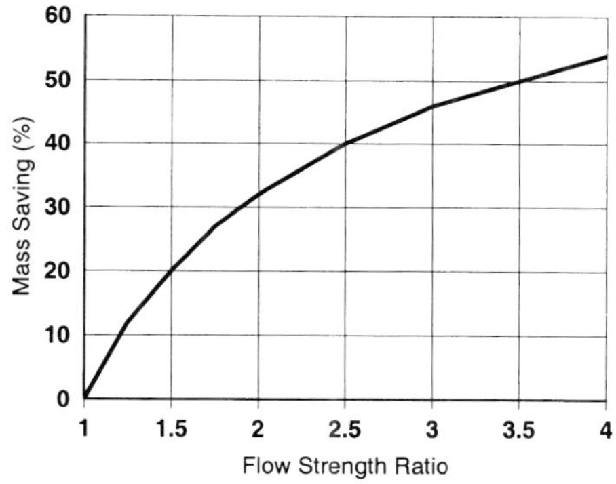

Figure 6: Mass saving as a function of the change in the flow strength (assuming t $^{1.8}$), after ref. [6].

It can be seen from Figure 6 that doubling the flow strength, will result in only a 30 % mass saving for an equivalent collapse resistance. It is therefore important to understand this non-linear relationship, and how it changes with all types of materials.

6. DYNAMIC STABILITY

To exploit the lightweighting potential of high strength steels the automotive designer needs to be able to downgauge components with confidence. This is especially important for crash management structures as the reduction in thickness has the potential to compromise dynamic stability.

For optimum performance in axial impact the retention of progressive collapse is required, as the transition to global buckling generally reduces energy absorption. This is substantially influenced by the ability of the supporting structure to carry the impact loads. In addition, the formation of poor folding modes may also initiate bending failure and such modes are typically a function of the component design, most especially the thickness to wall length ratio (t/b).

The reduction in sheet thickness therefore has the potential of reducing the t/b ratio to a critical level where poor non-compact folding modes would be initiated. However, high strength steels generally demonstrate increased dynamic stability. This is illustrated in Figure 7, which displays axially crushed square specimens manufactured from a mild steel and a high strength steel respectively.

The higher strength product collapsed progressively with tight energy efficient folds whereas the mild steel formed non-compact folding modes. These observations agree with other work [8] where high strength steels demonstrated significantly increased stability even in off-angle axial impact events.

High strength steels therefore not only provide significant increases in energy absorption but they also improve dynamic stability. This allows downgauging to be undertaken with confidence or alternatively for the wall length of the component to be increased with no compromise in the collapse efficiency.

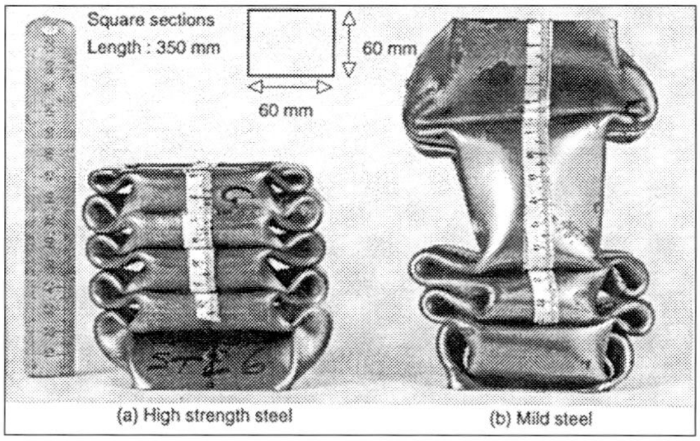

Figure 7. Quasi-statically tested sections manufactured from mild steel and a high strength material.

7. JOINING TECHNIQUE

The influence of dynamic stability demonstrates that effective energy absorption is a function of both the material properties and the component design. These two factors are interlinked and both need to be properly understood to maximise the lightweighting potential offered by high strength steel materials. An example of how the design can influence the collapse resistance, independently of the material, is in the application of continuous joining techniques, an example of which is laser welding.

Laser welding greatly increases the component stiffness [1], and can also be used within the flange of a structural member to influence its impact performance. This is shown in Figure 8.

Simple closed top hat structures were manufactured from mild steel, and the position of the laser weld within the flange varied. Laser welding in the middle of the flange greatly increased the collapse stability whilst laser welding at the root of the flange increased the collapse resistance, although slightly compromising the stability. In the latter case the collapse resistance was increased such that the mild steel structure exhibited similar performance to a spot welded closed top hat manufactured from high strength dual phase material.

This illustrates the possibilities of increasing energy absorption by understanding the behaviour of the structure under impact loading and the influence of the fabrication process.

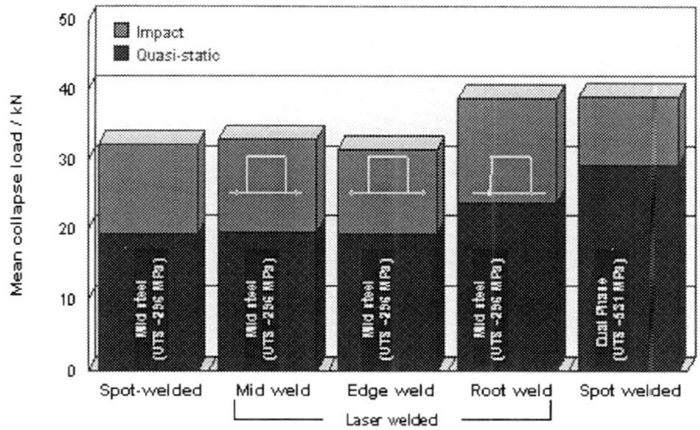

Figure 8. Influence of laser weld position on the collapse performance of closed top hat sections.

8. CONCLUSIONS

The mass saving potential of high strength steels in automotive components is significant. The identification of the important material properties together with an understanding of component design, the forming process and fabrication method allows their potential to be fully exploited.

The understanding of the relevant material properties for energy absorption is important for effective material selection in addition to facilitating the development of new grades of material especially designed for crashworthiness. Properties which are thought to contribute to energy absorption and provide effective weight reduction include the static strength, the strain rate sensitivity (upto perhaps 200 s^{-1}), significant static and dynamic work hardening and less obvious properties such as isotropic hardening behaviour. The response of such properties to the forming process is also important.

However, a better understanding of the relevance of current material selection criteria is required in reference to the actual deformation found in practice. Crash analysis plays an important role in facilitating this, and therefore, the material models need to be correctly defined. This will enable the utilisation of FE in an iterative process for both design and material selection.

The increased stability of high strength steels allows the designer to downgauge with confidence and the application of novel manufacturing methods such as hydroforming and the use of continuous joints may further enhance the lightweighting potential of such materials.

9. REFERENCES

1. Walker E. F. and Boyles M. W., 'Steel in lightweight automobiles - ULSAB and beyond', Third International Conf. on Materials for Leanweight Vehicles, Heritage Motor Centre, Gaydon, UK, 10–11th Nov., 1999.
2. Miura K. *et al.*, 'Dynamic deformation behaviour of steel sheet for automobile', SAE 960019, pp. 1–5, 1996
3. Miura K. *et al.*, 'High strain rate deformation of high strength sheet steels for automotive parts', SAE 980952, pp. 23–30, 1998.
4. Yoshitake A. *et al.,* 'Impact absorbed energy of hat square column in high strength steels', SAE 960020, pp. 7–15, 1996.
5. Mizui N. *et al.*, 'Fundamental study on improvement in frontal crashworthiness by application of high strength sheet steels', SAE 970156, pp. 39–44, 1997.
6. Marsh M.G. and McGregor I.J., 'The axial collapse performance of lightweight steel structures', Third International Conf. on Materials for Leanweight Vehicles, Heritage Motor Centre,Gaydon, UK, 10–11th Nov., pp. 53–64, 1999.
7. Sperle J.O. and Lundh H., 'Strength and crash resistance of structural members in high strength dual-phase steel sheet', Skand. J. of Metal., 13, pp. 343–351, 1984.
8. Nakanishi E. *et al.*, 'New materials technology for achieving both crashworthiness and weight reduction using energy-absorbing steel with higher strain rate sensitivity', SAE 980953, pp. 31–37, 1998.

